BRIDGE AND PIER PROTECTIVE SYSTEMS AND DEVICES

CIVIL ENGINEERING

A Series of Textbooks and Reference Books

Editors

ALFRED C. INGERSOLL
Associate Dean, Continuing Education
University of California, Los Angeles
Los Angeles, California

CONRAD P. HEINS, Jr.
Department of Civil Engineering
University of Maryland
College Park, Maryland

KENNETH N. DERUCHER
Consulting Engineer
Civil Design Corporation
Bowie, Maryland

BRIDGE AND PIER PROTECTIVE SYSTEMS AND DEVICES

KENNETH N. DERUCHER

Consulting Engineer
Civil Design Corporation
Bowie, Maryland

CONRAD P. HEINS, Jr.

Department of Civil Engineering
University of Maryland
College Park, Maryland

MARCEL DEKKER, INC.   New York and Basel

Library of Congress Cataloging in Publication Data

Derucher, Kenneth N.   [Date]
    Bridge and pier protective systems and devices.

    (Civil engineering ; 1)
    "Completed under U.S.C.G. contract CG-71955-A,
'A study of the state-of-the-art of bridge protective systems and devices.' "
    Includes bibliographical references and index.
    1. Bridges--Protection. 2. Fenders for
docks, piers, etc. I. Heins, Conrad P., joint
author. II. Title. III. Series.
TA1.C4523 vol. 1 [TG325] 624s [624.2] 79-16710
ISBN 0-8247-6895-7

This volume first appeared in November 1978, as a study
sponsored by the U.S. Coast Guard, under the title of
"The State-of-the-Art Bridge Protective Systems and
Devices."

MARCEL DEKKER, INC.

270 Madison Avenue, New York, New York 10016

Current printing (last digit)
10 9 8 7 6 5 4 3 2 1

PRINTED IN THE UNITED STATES OF AMERICA

PREFACE

Increasing demands are now being placed on our waterways and adjacent structures such as bridges, wharves, harbor piers, marinas, and lock entrances. These demands are attributable to the phenomenal growth in navigation modules whether it be a tanker, containership, or barge tow.

The increased number of collisions with bridges in recent years has focused more interest in the Coast Guard bridge replacement program. Designing a cost effective replacement bridge with a navigation prism, capable of meeting the present day needs of navigation, is easy to justify. These replacement bridges are highly cost effective because the replacement bridge provides for optimum productivity of navigation. In other words, vessels of economic size are able to utilize the waterway resources. However, the protection of bridge piers is another matter. They are not unlike the car bumper which provides no utilitarian value of space or mobility to the passenger.

Conceptually, protection of the vessels and bridge piers is provided by a fender system adjacent to the navigation opening of the bridge, rather than the vessel. The dilemma which now exists is, simply, that vessels have grown in magnitude and disproportionately to the growth in size or capability of the bridge protection systems. To completely protect against maximum possible impact of collision, in some cases, now dictates placement of a mass in the waterway so heavy that its cost exceeds the cost of the bridge it protects.

Such a mass, however, affords little or no protection to the vessel. These problems remain a concern of the bridge owner, the vessel operator, port authorities, and the Coast Guard. It certainly follows that full utilization of existing technology and identification of these areas of research is needed and will lead to better protective systems at lower cost.

Recognizing that a single source of reference, containing all existing technical data, methodology for selection of optimum fendering devices and recommended standards was urgently needed, the Coast Guard advertised for contract services for a State-of-the-Art Study for Bridge Protective Systems and Devices on 25 February 1977. The research contract was awarded to the Department of Civil Engineering, University of Maryland, on 15 July 1977 under the directions of Drs. K. N. Derucher and C. P. Heins.

This text shows the results of such a project. Chapter 1 is a brief introduction of the factors considered in design and the type of systems utilized. Chapter 2 discussed seven types of fendering systems as to their advantages and disadvantages. Chapter 3 deals with the material properties of the various fendering systems. Design parameters are discussed in much detail in Chapter 4 and the design of each by hand is shown in Chapter 5. Chapter 6 discusses the design applications and computers. Conclusions and Recommendations are dealt with in Chapters 8 and 9 respectively.

The text is intended for use as a guide or text for engineers, bridge owners, the vessel operators, port authorities, Coast Guard personnel, and educators. It is hoped that this text aids in the design of better and improved bridge protective systems and devices.

K. N. Derucher
C. P. Heins

ACKNOWLEDGMENTS

Special appreciation is extended to Mr. Michael P. O'Neill, Project Director, and Captain Richard D. Hodges, Chief, Bridge Division, for without their patience, time, and consideration this project would not have been accomplished.

We are also grateful for the assistance of Ralph T. Mancill, Jr., Chief, Bridge Modification Branch, U.S. Coast Guard.

This project was completed under U.S.C.G. Contract CG-71955-A, "A Study of the State-of-the-Art Of Bridge Protective Systems and Devices."

CONTENTS

TABLES

FIGURES

BRIDGE AND PIER PROTECTIVE SYSTEMS AND DEVICES

CHAPTER I

INTRODUCTION

GENERAL

Tankers, bulk carriers, cargo vessels and barges are being built increasingly larger in recent years. They require more room to maneuver and supply greater cargo capacities than ever before. This places increasing demands on waterways and adjacent structures such as bridges, docks, harbors, piers, marinas, lock and port entrances.

Coast Guard casualty statistics show that vessel collisions with fixed objects, such as bridges, more than doubled between 1966 and 1975 as larger and greater numbers of vessels used the nation's waterways. One Coast Guard study reveals that during the period FY71 and FY75, $23,153,000 in damage and fourteen fatalities were encountered. Obviously, such statistics indicate that a need exists to assure that proper design practices are used for fendering system installation. This need was recognized by the Bridge Division, U.S. Coast Guard, which is charged with the responsibility to provide for the economic efficiency and safety of marine transportation under bridges spanning the navigable waters of the United States. They have initiated a research contract with the Department of Civil Engineering, University of Maryland to conduct a study of the State-of-the-Art of Bridge Fendering Systems."

1

FACTORS CONSIDERED IN THE DESIGN

The function of bridge fendering systems are to protect bridge elements against damage from waterborne traffic. There are many factors to be considered in the design of fendering systems including the size, contours, speed, and direction of approach of the ship using the facility; the wind and tidal current conditions expected during the ship's maneuvers and while tied up to the berth; and the rigidity and energy-absorbing characteristics of the fendering system and ship, and finally the subgrade soil condition. The final design selected for the fender system will generally evolve after making arbitrary limitations to the values of some of these factors and after reviewing the relative cost of initial construction of the fendering system versus the cost of fender maintenance and of ship repair. In other words, it will become necessary to decide upon the most severe docking or approach conditions to protect against and design accordingly; hence, any situation which imposes conditions which are more critical than the established maximum would be considered in the realm of accidents and probably result in damage to the dock, fendering system (whether used for dock or approach conditions) or the ship.

TYPES OF FENDERING SYSTEMS

As a result of the factors considered in the design, many fendering systems have been designed and/or analyzed. These systems are of wide variety and material which vary considerably in design, fabrication, and cost. As a result of the literature search, basically seven types of fendering systems are in existence. These seven systems are as follows:

1. Floating Fender or Camels

2. Standard Pile-Fender System

   a. Timber-Pile

   b. Hung Timber

   c. Steel Pile

   d. Concrete Pile

3. Retractable Fender System

4. Rubber Fender System

   a. Rubber in Compression (Seike)

   b. Rubber in Shear (Raykin)

   c. Lord Flexible

   d. Rubber in Tension

   e. Pneumatic

5. Gravity Type Fender System

6. Hydraulic and Hydraulic-Pneumatic Fender System

   a. "Dashpor" Hydraulic

   b. Hydraulic-Pneumatic Floating Fender

7. Spring Type Fender System

As indicated, the above seven (7) fendering systems were found to be in existence throughout the world by a comprehensive library and computer literature search. These systems will be explained in subsequent chapters as to advantages, disadvantages, design parameters, materials of construction and design procedures (hand and computer oriented).

CHAPTER II

TYPES OF FENDERING SYSTEMS

This chapter deals with the various types of fendering systems as mentioned in Chapter I. The uses, advantages, and disadvantages of each type will be discussed. Although the authors discuss the fendering systems in general terms, it should be remembered that fenders are applicable to both the docking and mooring problem as well as bridge protection.

FLOATING FENDER OR CAMEL

The floating fender or camel is the simplest type of fendering system employed. This is often used in combination with other types of fendering to provide a large contact area and thereby increase the efficiency of the fendering system while decreasing the load concentration at any one point on the fender. They may be used as rubbing surfaces to protect the ship's hull from damage by the dock structure during berthing procedures or from damage by the action of other ships while moored. Traditionally, the camel has consisted of treated logs of Douglas fir, Southern pine, White oak, Red oak, or any wood with relatively high fiber hardness and high bending strength. Greenheart and other exotic woods have natural marine borer resistance but are heavier than water, making their use as camels impractical in most cases. The treated timber is hung on chains in single units, bundled groups, or dapped and bolted heavy box sections. The high weight, strength, and stiffness of the box sections have limited their use to busy Navy installations where they have performed satisfactorily in most instances. Energy absorption occurs as a result of crushing of the timber fibers and bending of

the camel groups. Loss of strength may occur where the timber has been
splintered due to faulty construction practices or high energy considerations or where degradation of the wood has occurred due to marine borer attack or rotting. Though the camel has been used for years as a component
of many fendering systems, it may result in severe damage to the ship's hull
and fender, dock or bridge structure as a result of concentrated loadings
along the length of the camel during high energy impacts. Projecting hard-
ware on the camel or fendering structure may result in tears or punctures
to the ship's hull.

The latest development in floating fenders came in 1967 in which the
Hi-Dro Cushion camel (Fig. 1) was constructed and tested at Treasure Island
Naval Station in San Francisco, California. The Hi-Dro Chsion camel consists of eighty-four, three foot water filled cells, grouped into four clusters, sandwiched between two timber rubbing faces and held in place by cables.
The water filled cylinders maintain constant pressure during compression upon
impact by forcing water out through small orifices in the tops of the cylinders. The compression of the cylinders and the resulting hydraulic action,
as well as the bending of the timber faces, crushing of the timber fibers,
and movement of the water between the cell clusters, result in higher energy
absorption than that of the traditional camel. In addition, the Hi-Dro Cushion can adjust to varying ship displacement, approach velocities, and environmental factors.

STANDARD PILE FENDER SYSTEM

The timber-pile (Fig. 2) system employs piles driven along a wharf, face
bottom. Pile tops may be unsupported laterally or supported at various degrees of fixity by means of whalers and chocks. Single or multiple row
whalers may be used, depending on pile length and tidal variation. Impact

5

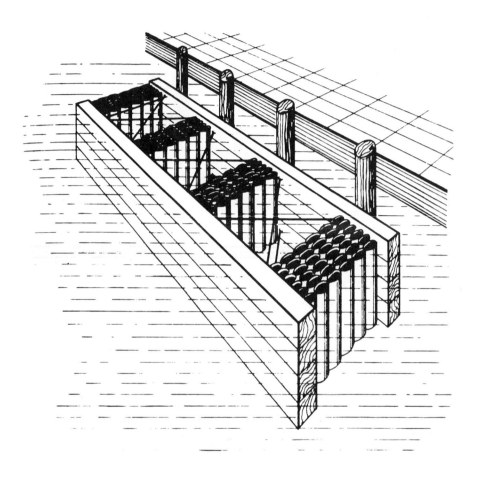

FIG. 1: Hi-Dro Cushion Camel (Courtesy of Rich Enterprises)

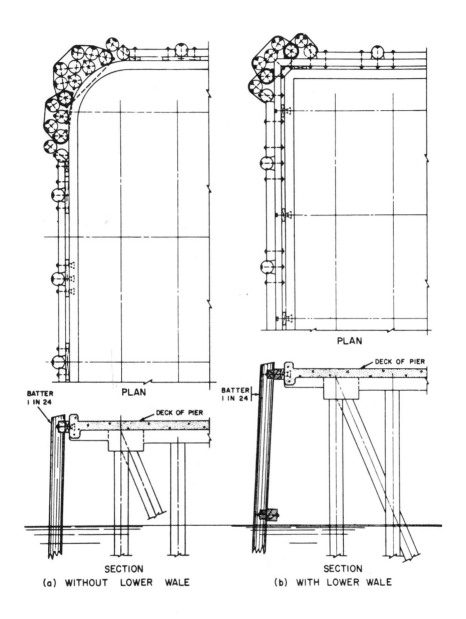

FIG. 2: Timber Fender Piles (Courtesy of U.S. Navy)

energy upon a timber fender pile is absorbed by deflection and the limited compression of the pile. Timber piles are abundant and have a low initial cost. They are susceptible to mechanical damage and biological deterioration. Once this happens, the energy-absorption capacity declines and a high maintenance or repair cost results. Timber piling can be observed in satisfactory service in most marinas where smaller recreational vessels are berthed and moored. For large vessels and less sheltered waterways, steel, concrete, or composite sections have gained wide acceptance.

The hung timber system (Fig. 3) consists of timber members fastened rigidly to the face of the dock. A contact frame is formed which distributes impact loads, but its energy-absorption capacity is limited and it is unsuitable for locations with significant tide and current effects. Whalers should be provided along the length of the rubbing strip with additional whalers provided near mean low and mean high water. The hung timber system has a low initial cost and is a less bio-deterioration hazard than the standard timber pile.

Since World War II various materials have been used in conjunction with the hung timber fenders. Cylindrical and rectangular rubber pieces, hydraulic units, and steel springs have been inserted behind the timber framework in order to improve the energy absorption characteristics of the hung fender.

Steel fender piles are occasionally used in water depths greater than forty feet, or for locations where very high strength is required and a difficult seafloor condition results. Its main disadvantages are high cost and its vulnerability to corrosion.

Regular, reinforced concrete piles are not satisfactory because of their limited internal strain-energy capacity and the steel reinforcement may corrode due to concrete cracking. Prestressed concrete piles with rubber buffers at deck level have been used. In this case, the rubber units are the principa

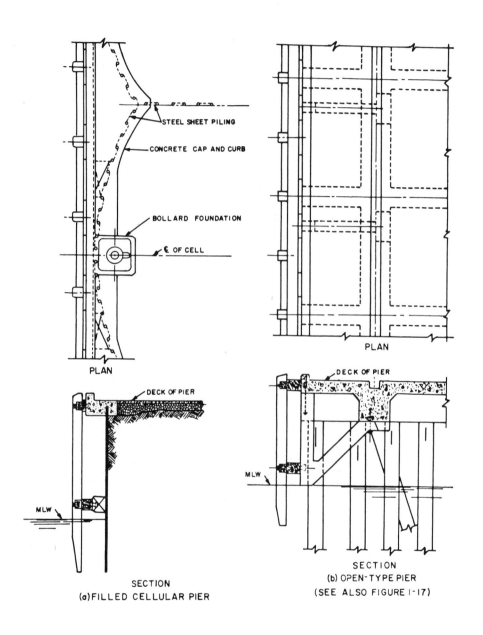

FIG. 3: Hung Timber Fenders (Courtesy of U.S. Navy)

energy-absorbing media; and not the piles. This system is very resistant to natural and biological deterioration.

RETRACTABLE FENDER SYSTEM

Retractable fenders generally consist of a timber or steel framework with timber facing members supported on open pins or slotted brackets (Fig. 4) fastened to the pier or dock structure. Energy absorption occurs as a result of the vertical displacement of the fender, friction in the brackets, and bending and crushing of the timber fibers. When berthing maneuvers result in an impact force greater than the resultant of the weight of the fender in the direction of the brackets and the friction of the brackets, the fender begins to move upward and backward toward the dock face. As this motion occurs, more timber facing strips are brought into contact with the ship's hull increasing the resistance of the fender to further movement. Impact energies are rarely of sufficient magnitude to push the fender to the dock face, but, should this occur, additional fendering is provided by the deformation of the timber facings and then the underlying framework.

Design considerations must include the effective weight of the fendering system which is directly affected by tidal variations, maximum retracting distance, angle of inclinations of the brackets, and the sufficient number of brackets. Damage to the supporting brackets can render the fender inoperative although replacement of the timber facing strips does not adversely affect energy absorption or present difficulties for repair. Provisions for glancing blows may include allowance of a slight amount of slip in the brackets. The fender will then bear upon the bracket wall of each fendering section which comes in contact with the vessel.

Recent innovations in the area of retractable fenders include the use of steel framework with rubber tires or cylindrical fenders positioned around

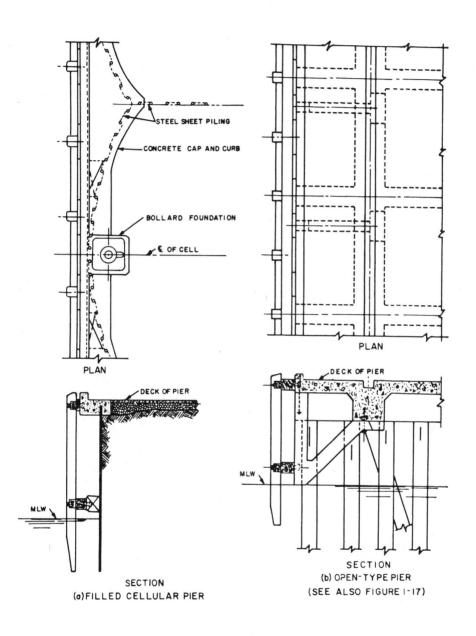

STEEL SHEET PILING

CONCRETE CAP AND CURB

BOLLARD FOUNDATION

₵ OF CELL

PLAN

PLAN

DECK OF PIER

DECK OF PIER

MLW

MLW

SECTION
(a) FILLED CELLULAR PIER

SECTION
(b) OPEN-TYPE PIER
(SEE ALSO FIGURE I-17)

FIG. 3: Hung Timber Fenders (Courtesy of U.S. Navy)

9

energy-absorbing media; and not the piles. This system is very resistant to natural and biological deterioration.

RETRACTABLE FENDER SYSTEM

Retractable fenders generally consist of a timber or steel framework with timber facing members supported on open pins or slotted brackets (Fig. 4) fastened to the pier or dock structure. Energy absorption occurs as a result of the vertical displacement of the fender, friction in the brackets, and bending and crushing of the timber fibers. When berthing maneuvers result in an impact force greater than the resultant of the weight of the fender in the direction of the brackets and the friction of the brackets, the fender begins to move upward and backward toward the dock face. As this motion occurs, more timber facing strips are brought into contact with the ship's hull increasing the resistance of the fender to further movement. Impact energies are rarely of sufficient magnitude to push the fender to the dock face, but, should this occur, additional fendering is provided by the deformation of the timber facings and then the underlying framework.

Design considerations must include the effective weight of the fendering system which is directly affected by tidal variations, maximum retracting distance, angle of inclinations of the brackets, and the sufficient number of brackets. Damage to the supporting brackets can render the fender inoperative although replacement of the timber facing strips does not adversely affect energy absorption or present difficulties for repair. Provisions for glancing blows may include allowance of a slight amount of slip in the brackets. The fender will then bear upon the bracket wall of each fendering section which comes in contact with the vessel.

Recent innovations in the area of retractable fenders include the use of steel framework with rubber tires or cylindrical fenders positioned around

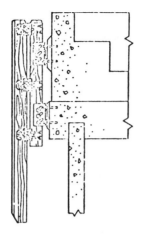

(a) Slotted supporting brackets

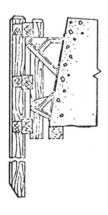

(b) Open pin supporting brackets

FIG. 4: Retractable Fender Brackets (Courtesy of U.S. Navy)

the vertical axis allowing free rotation about the vertical axis. This system incorporated counterweights to counteract the negative moment of portions of the frame loaded seaward from the bearing piles. The rubber tires eliminate many of the concerns regarding glancing blows by allowing free rotation, decreasing the severity of the reaction force which would result from timber sections, and reduce wear and maintenance of contact members.

RUBBER FENDER SYSTEMS

The rubber fender systems occur in five types as outlined earlier. Rubber in compression (Seike) consists of a series of rubber cylindrical or rectangular tubes installed behind standard fender piles or behind hung-type fenders. Energy absorption occurs as a result of the deflection of the fender and/or the internal stress-strain characteristics of the rubber material. When bolted directly to the face of the pier, care must be taken to insure that all bolts are recessed and firmly anchored in the dock structure (Fig. 5). In design, a proper bearing timber-frame is required for transmission of impact forces from ship to pier.

Cylindrical and rectangular units can be used interchangeably in most instances (Fig. 6). When mounted on the flat face of a pier, they may be bolted directly to the face of the dock, draped, hung vertically (Figs. 7 and 9), or layered using one or more of these methods (Fig. 8). On curved structures, draping is not possible and the fenders must be bolted to the face of the pier or hung vertically. Draped fenders require a solid dock face of at least six feet to insure at least a three foot depth of contact between the fender and the ship's hull. This makes pre-curved sections a necessity. All connections should be recessed so that they do not protrude beyond the face of the dock. The low point of each fender should be fitted with a drain hole.

FIG. 5: Rectangular Rubber Fender (Courtesy of Goodyear)

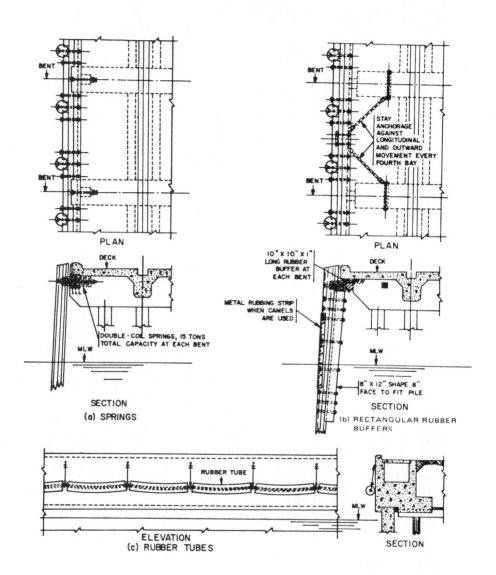

BENT

BENT

PLAN

DECK

DOUBLE-COIL SPRINGS, 15 TONS
TOTAL CAPACITY AT EACH BENT

MLW

SECTION
(a) SPRINGS

BENT

STAY
ANCHORAGE
AGAINST
LONGITUDINAL
AND OUTWARD
MOVEMENT EVERY
FOURTH BAY

BENT

PLAN

10" X 10" X 1"
LONG RUBBER
BUFFER AT
EACH BENT

DECK

METAL RUBBING STRIP
WHEN CAMELS ARE USED

MLW

8" X 12" SHAPE 8"
FACE TO FIT PILE

SECTION
(b) RECTANGULAR RUBBER
BUFFERS

RUBBER TUBE

MLW

ELEVATION
(c) RUBBER TUBES

SECTION

FIG. 6: Mounting of Rubber Fenders (Courtesy of U.S. Navy)

14

a. Cylindrical

b. Trapezoidal

FIG. 7: Rubber Fenders (Courtesy of Goodyear)

FIG. 8: Layered Rubber Fenders (Courtesy of Goodyear)

FIG. 9: Draped Rectangular Rubber Fenders (Courtesy of Uniroyal)

When rubber fenders are applied to the dock face in any of the above manners, certain problems arise due to the contact of the vessel and rubber fender. The rubbing action of the ship's hull against the fender may damage the fender if unevenness exists along the hull due to rivets, butt joints, or any number of slightly protruding elements.

Energy absorption capacity of such a system can be varied by using the tubes in single or double layers, or by varying tube sizes. The energy absorption of a cylindrical tube is nearly directly proportional to the ship's force until the deflection equals approximately one-half the external diameter; after that, the force increases much more rapidly than the absorption of energy.

Another method of utilizing rubber in compression is to mount the rubber behind timber rubbing strips thereby eliminating the contact of the rubber against the ship and all of the ensuing difficulties (Figs. 10 and 11). In this manner, additional energy absorption is provided by the bending and crushing of the timber fibers. Rubber fenders may be positioned with the bore parallel or perpendicular to the dock face. When positioned perpendicular to the dock so that the sections are axially end-loaded (Fig. 12), the energy absorption and reaction force curves are very nearly linear. When positioned parallel to the dock so that the sections are loaded perpendicular to the bore of the fender, there is a disproportionate increase in both the energy and force for a given deflection (Fig. 13). Therefore, using either the cylindrical or rectangular fenders in axial compression allows design over a larger range of deflections than using either in radial compression.

Still, another type of rubber fender used in compression is the trapezoidal fender of which two types exist: the internal bottom mounting plate (Fig. 14a) and the external bottom mounting plate (Fig. 14b). These fenders

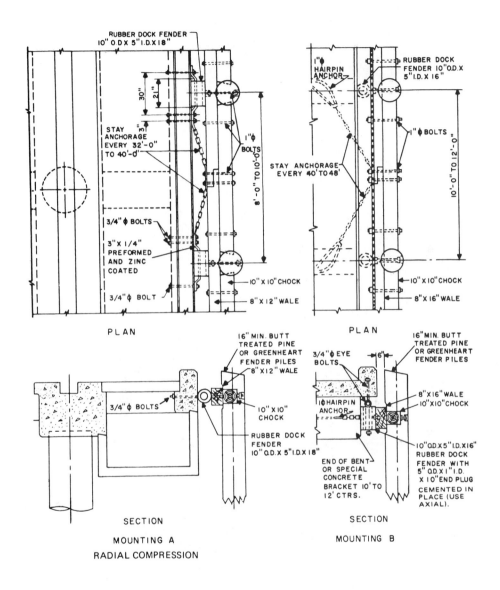

FIG. 10: Rubber Fenders Mounted Behind Hung Timber Fenders
(Courtesy of Little)

FIG. 11: Rubber Fenders Mounted Between Wood Assemblies
(Courtesy of Uniroyal)

FIG. 12: Axially Mounted Rubber Fenders (Courtesy of Goodyear)

FIG. 13: Side Loaded Rubber Fender Assemblies (Courtesy of Uniroyal)

22

a. Vertically Mounted External Plate

b. Horizontally Mounted Internal Plate

FIG. 14: Trapezoidal Rubber Fenders (Courtesy of Goodyear)

may be bolted either horizontally or vertically to the face of a dock or they may be applied behind timber rubbing strips. When mounted between the timber facing and the dock, they may be used singly or may be bolted together to form what resembles a large rubber "X" either with or without a top or bottom (Fig. 15).

Rubber funders in compression may also be applied directly to the vessels (Fig. 16).

The Raykin fender (Fig. 17) is the most commonly used rubber in shear system. The Raykin consists of a series of rubber pads bonded between steel plates to form a series of "sandwiches" mounted firmly as buffers between a pile-fender system and pier. Two types of mounting units are available, which are capable of absorbing 100 percent of the energy. The only problem with the rubber in shear fenders is that they tend to be too stiff for small vessels and the steel plates have a tendency to corrode. Therefore, it follows that they have a high energy absorbing capacity for larger ships.

Solid cylinders and rectangular rubber members have been used as shear fenders. By attaching the rubber section to the underside of the dock and to a horizontal piece of timber or steel to which timber rubbing strips are attached, the rubber is put into direct shear by any impact load (Figs. 18 and 19). As with the Raykin fender, this system is best suited for a large, uniform ship size and is not soft enough for smaller ships.

The Lord flexible fender (Fig. 20) consists of an arch-shaped rubber block bonded between two end steel plates. Under axial loading, the Lord fender buckles in much the same way that a column will buckle if the slenderness ratio becomes large enough to cause instability. It can be installed on open or bulkhead type piers and on dolphins, or incorporated with standard piles as the hung system. Impact energy is absorbed by bending and compression of the arch-shaped rubber column. With the Lord flexible fender, possible destruction of the bond between the steel plates and rubber may result.

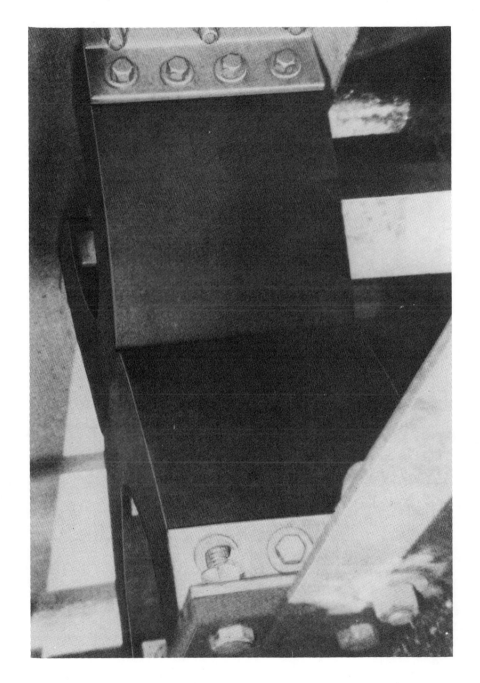

FIG. 15: Double Mounted Trapezoidal Rubber Fenders (Courtesy of Uniroyal)

FIG. 16: Rubber Fenders Aboard Ship (Courtesy of Goodyear)

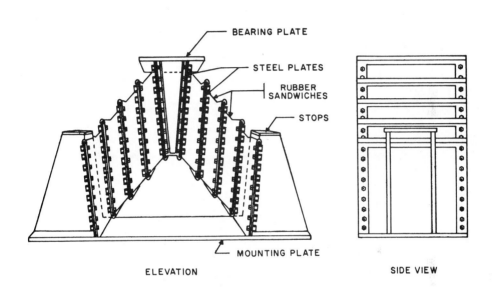

FIG. 17: Raykin Fender (Courtesy of U.S. Navy)

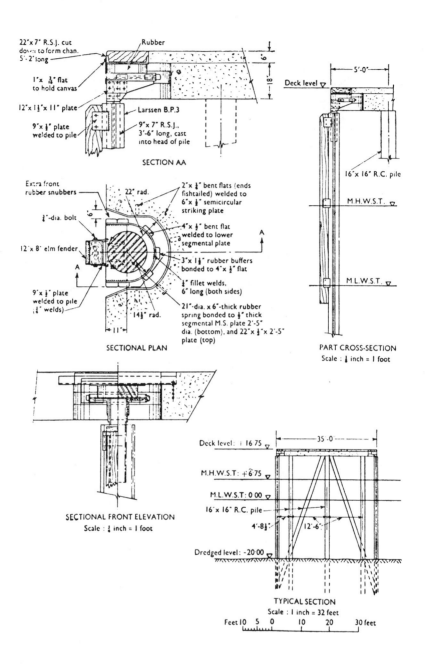

SECTION AA

- 22"x 7" R.S.J. cut down to form chan. 5'-2" long
- Rubber
- 1"x ⅛" flat to hold canvas
- 12"x 1½"x 11" plate
- 9"x ½" plate welded to pile
- Larssen B.P.3
- 9"x 7" R.S.J., 3'-6" long, cast into head of pile
- 6"
- 18"

SECTIONAL PLAN

- Extra front rubber snubbers
- 22" rad.
- ¾"-dia. bolt
- 6"
- 12"x 8" elm fender
- 9"x ½" plate welded to pile (¼" welds)
- 11"
- 14½" rad.
- 2"x ¼" bent flats (ends fishtailed) welded to 6"x ½" semicircular striking plate
- 4"x ½" bent flat welded to lower segmental plate
- 3"x 1½" rubber buffers bonded to 4"x ½" flat
- ¼" fillet welds, 6" long (both sides)
- 21"-dia. x 6"-thick rubber spring bonded to ½" thick segmental M.S. plate 2'-5" dia. (bottom), and 22"x ½"x 2'-5" plate (top)
- A   A

SECTIONAL FRONT ELEVATION
Scale : ¼ inch = 1 foot

PART CROSS-SECTION
Scale : ¼ inch = 1 foot

- Deck level ▽
- 5'-0"
- 16"x 16" R.C. pile
- M.H.W.S.T. ▽
- M.L.W.S.T. ▽

TYPICAL SECTION
Scale : 1 inch = 32 feet

- Deck level: + 16.75 ▽
- 35.0
- M.H.W.S.T: +6.75 ▽
- M.L.W.S.T: 0.00 ▽
- 16"x 16" R.C. pile
- 4'-8½"
- 12'-6"
- Dredged level: -20.00 ▽

Feet 10 5 0   10   20   30 feet

FIG. 18: Rubber Fender in Shear (Courtesy of Little)

28

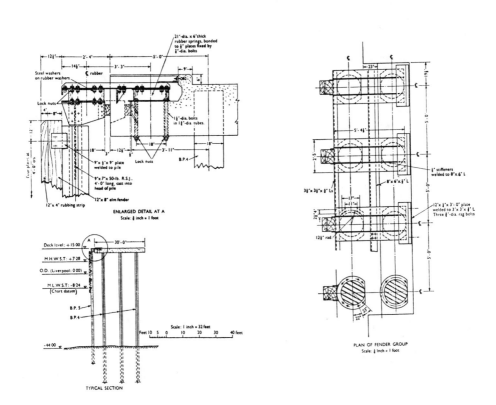

FIG. 19: Rubber Fender in Shear Mounting (Courtesy of Little)

29

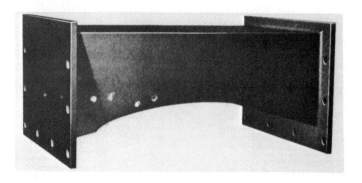

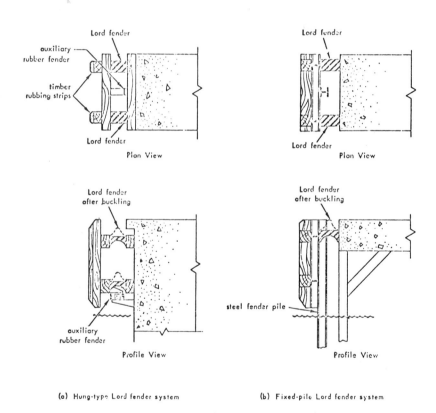

(a) Hung-type Lord fender system   (b) Fixed-pile Lord fender system

FIG. 20: Lord Flexible Fender (Courtesy Lord Co. and U.S. Navy)

Rubber in tension fenders consists of a combination of rubber and steel fabricated in a cone-shaped, compact bumper form, molded into a specially cast steel frame, and bonded to the steel. It absorbs energy by torsion, compression, shear, and tension. Their main disadvantage is possible destruction of the bond between steel castings and the rubber.

The final category of rubber fendering systems is the pneumatic fenders. These are pressurized, airtight rubber devices designed to absorb energy by compression of air inside a rubber envelope (Fig. 21). Pneumatic fenders have recently been applied to fixed dock-fender systems (Fig. 22) and are feasible for use as ship fenders or shock absorbers on floating fender systems. Two pressure ranges are used in the pneumatic fenders, approximately one psi and seven psi. A proven fender of this type is the pneumatic tire wheel fender which consists of pneumatic tires and wheels capable of rotating freely around a fixed or floating axis. The fixed unit is designed for incorporation in concrete bulkheads. The floating unit may consist of two to five tires. Energy-absorption capacity and resistance load depends on the size and number of tires used and on initial air pressure when inflated.

Pneumatic fenders can be hung by cable or chains through the metal end rings which are located at each end or they may be encased in a net and hung at various points along the fender length (Fig. 23). When used on a dock or pier, the abrasion resulting from berthing and mooring can be prevented by the use of tires fastened to the pneumatics by a network of rope or cable. The range of sizes available allows for design of an adequate fendering system even when space is critical.

A recent development in the pneumatic field is the use of foam filled fenders (Fig. 24). During accidental collisions, there is a slight chance that the pneumatic fenders may be overloaded and release air through a release valve or through a puncture. In this event the fender becomes completely

31

FIG. 21: Pneumatic Fenders in Whaling (Courtesy of Yokohama)

FIG. 22: Pneumatic Fenders (Courtesy of Yokohama)

33

FIG. 23: Anchoring for Pneumatic Fender (Courtesy of Seward)

FIG. 24: Foam Filled Fenders (Courtesy of Dunlop)

inoperable. Although repair may be possible on site, the fender is completely out of commission and the dock is unprotected for the length of time necessary to effect repairs. The foam filled fender is unsinkable and, when punctured, will remain afloat and operative until removed for repair. Foam filled and pneumatic fenders have few maintenance problems. Aside from punctures or other accidental damage, the only maintenance required are occasional checks of the fenders themselves and their supporting chains or cables.

GRAVITY FENDERS

Gravity fenders (Fig. 25) consist of a large weight, generally concrete, suspended by chains or cables from beneath a dock structure. Wood facings transfer impact energy from the ship's hull to the suspended weight. The displacement of this weight is the main energy absorption mechanism with additional energy provided by the crushing of timber fibers or deflection of rubber if present. High energy absorption is achieved by long travel of the weight. Due to the size of the gravity fender, it is best suited for exposed locations where the vessel size is large. The rubbing strips must have sufficient "give" to allow the ship to act on the concrete block without damage to the ship's hull. Among the difficulties of the gravity fender are the construction and repair problems inherent in its large size. Heavy equipment is required for both installation and replacement. The suspension system is subjected to not only mechanical wear, but also sea water corrosion and must be maintained. These problems result in both high initial and high maintenance costs. The dock structure required to support the gravity fender may also be expensive to design and construct. The swinging action of gravity fenders has resulted in serious damage to the supporting structure when glancing blows or accidental collisions occur.

HYDRAULIC AND HYDRAULIC-PNEUMATIC FENDER SYSTEM

Hydraulic fenders are of two types: hydraulic-dashpot and hydraulic-pneumatic (Fig. 24). The dashpot consists of a fluid filled cylinder fitted with a plunger which when depressed, forces fluid through a variable of non-variable orifice. The fluid is generally forced to an elevated reservoir so that gravity and the high pressure return the dashpot to its original position once the impact energy is dissipated. The variable orifice changes its cross-sectional area with respect to the plunger travel thereby maintaining a constant reaction force. This system is similar to that used in railroad cars to prevent damage from longitudinal blows. The hydraulic unit is mounted between the dock structure and a flexible fender. The advantages of this system include nearly 90 percent efficiency in dissipation of kinetic energy. The hydraulic system maintains this efficiency over a wide range of velocities, maintains a constant force, reacts with minimum rebound resulting in little or no drift. The adaptability of these fenders allows use in a wide range of conditions by changing the metering system without substituting larger units. Corrosion or severe impact loads may result in serious damage to the unit rendering it inoperative. Repair can usually be effected without heavy equipment but during the repair period, the only fendering available is that provided by the piles or timber rubbing strips.

Hydraulic-pneumatics are a hybrid form of the pneumatic and hydraulic fenders. A rubber envelope is filled with water or air and water. Energy absorption occurs as a result of viscous resistance and air compression. These fenders are suspended in the same manner as the air-filled and the foam-filled pneumatics and share similar abrasion characteristics.

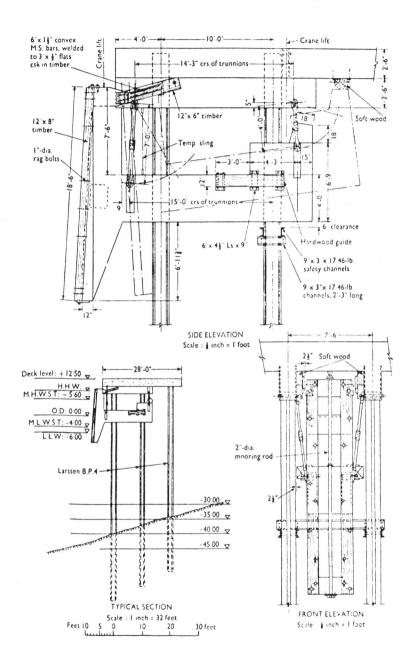

FIG. 25: Gravity Fender (Courtesy of Little)

SPRING TYPE FENDER SYSTEM

Steel spring fenders consist of steel springs mounted between the dock structure and a flexible fendering system, either piles or timber frameworks. The most common configuration consists of steel piles framed together with a timber facing of some sort bearing on steel plungers and steel springs. The springs are housed in a metal box with a removable cover to provide easy maintenance and repair of the springs. Maintenance of the springs involves coating, cleaning and corrosion checks. The principle energy absorption mechanism is the compression of the spring with additional energy provided by the deflection of the piles or timber framework. Although some of the installations where steel springs have been employed have provided adequate service, others have experienced difficulty due to failure of the spring as a result of corrosion. The nature of the spring assembly makes this fendering system suitable for exposed sea location and a uniform ship size.

RESULTS AND DISCUSSIONS

The advantages and disadvantages of each type of fendering system as well as its method of operation was explained in detail. These fendering systems as described are utilized in one form or another worldwide. The majority are of vintage years and outdated.

The design methods used for each system are based on the energy-absorption method in which no emphasis is placed on distribution factors. Further, as the ships increase in size and velocity, it has been the trend to increase the size of the fendering system rather than to look for new methods of design or for new systems.

Research at present by the University of Maryland is directed towards an improvement in design techniques and the development of new fendering systems which can satisfactorily handle larger vessels and speeds.

CHAPTER III

MATERIALS

Materials used as bridge protective devices consist of three principle
classifications: wood, steel, and concrete. This chapter is concerned with
these classifications as they are related to deterioration.

WOOD

Maritime structure constructed of wood have a long, colorful history.
While some structures have been an unqualified success, others have met with
dismal failure. Those factors which cause damage to the wooden structure
shall be discussed in the following order: fungi attack, insect attack, fire
damage, chemical damage, impact and overload failures. The various types of
wood will also be discussed emphasizing the possible advantages of hardwoods
as opposed to softwoods.

Molds, stains, and decay in timber are caused by fungi, microscopic
plants requiring organic material on which to live. Reproduction occurs
through thousands of small windblown particles called spores. Spores send
out small arms which destroy timber through the action of enzymes. Fungi require a temperature of between 50°-90° F., food and moisture. Timber which
is water soaked or dry normally will not decay. Molds cause little direct
staining because the color caused is largely superficial. The cottony or
powdery surface growths range in color from white to black and are easily
brushed or planed off the wood. Stains, on the other hand, cannot be removed
by surface techniques. They appear as specks, spots, streaks, or patches of
varying shades and colors depending upon the organism which is infecting the

timber. While stains and molds should not be considered as stages of decay since the fungi do not attack the wood substance in any great degree, timber infected with mold and stain fungi has a greater capacity to absorb water and is, therefore, more susceptible to decay fungi. Decay fungi may attack any part of the timber causing a fluffy surface condition indicative of decay or rot. Early stages of decay may show signs of discoloration or mushroom-type growths (Fig. 26). In some types the color differs only slightly from the normal color giving the appearance of being water soaked. Later stages of decay are easily recognized because of the definite change in color and physical properties of the timber (Fig. 27). The surface becomes spongy, stringy, or crumbly, weak, and highly absorbant. It is further characterized by a lack of resonance when struck with a hammer. Brown, crumbly rot in a dry condition is known as "dry rot". Serious decay problems are indicative of faulty design or construction or a lack of reasonable care in the handling of the timber. Principles that assure long service life and avoid decay hazards in construction include building with dry timber, using designs that will keep the wood dry and accelerate rain runoff, and using preservative-treated wood where the wood must be in contact with water.

Insect damage may result from infestation of the timber members with any of a number of insects: powder-post beetles, termites, marine borers, etc. Powder-post beetles are reddish-brown to black, hard-shelled insects from 1/8 to 1/2 inches long. The life cycle of the beetle includes four distinct stages: egg, larva, transformation, and adult. The adult bores into the timber producing a cylindrical tunnel just under the surface in which the eggs are laid. The larva burrow through the wood leaving tunnels packed with a fine powder 1/16 to 1/8 inch in diameter. Powder-post damage is indicated by this fine powder either fallen from the timber or packed into tunnels within the wood and by the tunnels and holes in the timber. The beetles

41

FIG. 26: Early Stages of Decay at Base of Timber Member

FIG. 27: Advanced Stage of Decay

attack both sound and decayed wood, but are not active in decayed wood which is water soaked. They may also cause damage by transmitting destructive fungi from one site to another, thus spreading decay.

Termites resemble ants in size and general appearance and live in similarly organized colonies. Destruction is done by the workers only, not by the soldiers or winged sexual adults. Subterranean termites are responsible for most of the damage done in the United States. They are more prevalent in the southern states, but are found in varying numbers throughout the states. Termites build dark, damp tunnels well below the surface of the ground with some of these tunnels leading to the wood which is the termites' source of food. Termites also require a constant water supply in order to survive. Subterranean termites do not infest structures by being carried to the construction site. They must establish a colony in the soil before they are able to attack the timber. Tell-tale signs are the tunnels in the earth leading to unprotected timber and swarms of male and female winged adults in the early spring and fall (Fig. 28). When termites successfully enter the wood, they make tunnels which follow the wood grain, leaving a shell of sound wood to conceal the tunnels (Fig. 29). Methods of controlling termites include breaking the path from the timber to the ground although the best method is to treat the timber with a preservative.

Wood inhabiting termites are found in a narrow strip around the southern boundary of the United States. They are most common in southern California and southern Florida. They are fewer in number than the subterranean termites, do not multiply as rapidly, and do not cause as much damage. But because they can live without contact with the ground and in either damp or dry wood, they are a definite problem and do considerable damage in the coastal states. They are carried to a building site in timber that has been infested prior to delivery, thereby making inspection prior to delivery a necessity. Full length treatment with a good wood preservative is recommended.

44

FIG. 28: Swarming Termites

45

FIG. 29: Characteristic Termite Tunnels

Carpenter ants are usually found in stumps, trees, or logs, but are some-times found in structural timbers. They range in color from brown to black and in size from large to small. Though carpenter ants are often confused with termites, they can be distinguished by caomparison of the wings and thorax (waist) sizes of the insects. The carpenter ant has short wings and a narrow waist, whereas the termite has long wings and a thicker waist. Car-penter ants use wood as shelter, not food. They prefer naturally soft or de-cayed wood and construct tunnels which are very smooth and free of dust. Car-penter ants tunnel across the wood grain and cut small exterior openings for access to their food supplies (Fig. 30). A large colony takes from three to six years to develop. Prevention of ant infestation can be accomplished by the use of preservatives along the length of the timber member.

By far the most damaging insect pest is the marine borer (Fig. 31). Borers have been known to ruin piles and framing within a few months. No ocean is completely free of borers and, while some waters may be relatively free, the status of an area may change drastically within a relatively short period of time. The main point of attack is generally between the high tide level and the mud line. Submergence often hides tell-tale signs of infesta-tion so that the first sign of attack may be the failure of the structure. The borers which do the most damage are the mollusks, related to clams and oysters, and the crustaceans, related to lobsters and crabs.

The mollusk borers consist of the "shipworm" teredo, the "shipworm" bankia, and the Pholadidae. There are other types of shipworms throughout the world, but they all live and survive in much the same ways though size and environmental requirements may vary. The teredo has a wormlike, slimy, gray body with two shells attached to the head which are used for boring. Two tubes that resemble a forked tail and normally remain outside the burrow are the only external indications that a shipworm is present. The shipworm

47

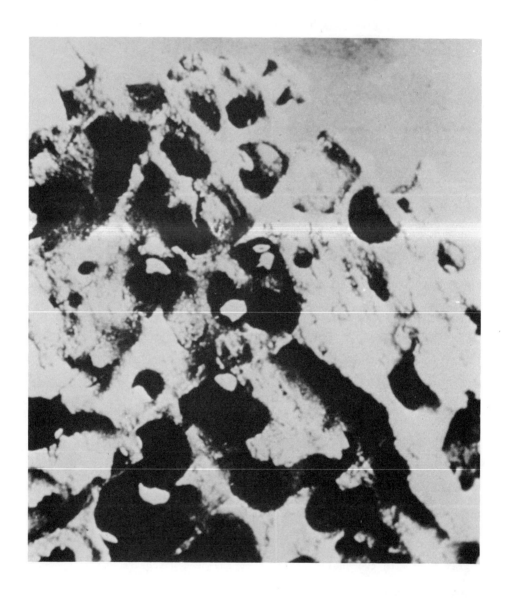

FIG. 30: Carpenter Ant Tunnels

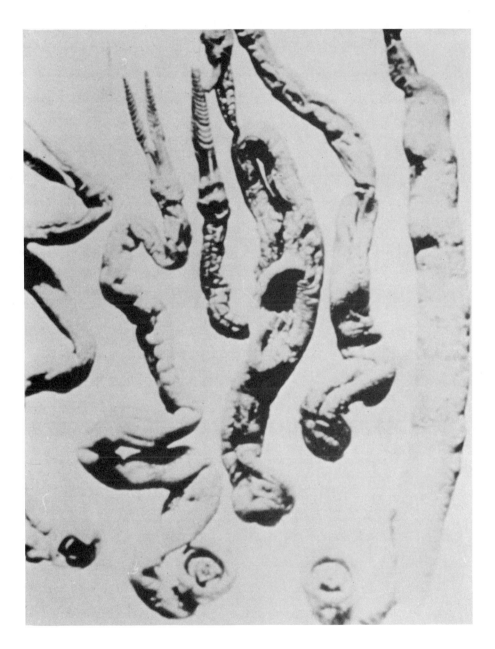

FIG. 31: Marine Borers

49

can seal its entrance and thereby protect the burrow from intruders or foreign substances. The size of the teredo varies from 3/8 to 1 inch in diameter and from six inches to six feet in length. The size of the bankia is generally larger than the teredo while other characteristics remain the same. Both types bore tiny holes when they are young and grow to maturity inside the wood. Once the young shipworm has entered the wood, it normally turns down and expands its burrow to its full diameter. Extremely careful observation with a hand lens is required to detect the entrance. The only way to detect the extent of the damage is to cut the wood (Fig. 32) or take borings by some other method. Because the first sign of marine borer infestation may be the failure of timber members (Fig. 33), the shipworm should be considered extremely dangerous. Pholadidae resembles a clam with its body entirely enclosed in a two-part shell. It is a particular danger because it can bore holes up to 1-1/2 inches deep into the hardest timber.

The most common crustacean borer is the limnora or "wood louse". Its body is slipper-shaped from 1/8 to 1/4 inch long and from 1/16 to 1/8 inches wide. It has a hard boring mouth, two sets of antennae, and seven sets of legs with sharp claws. The limnora is able to roll itself into a ball, to swim and to crawl. It will gnaw interlacing branching holes in the surface of the wood with as many as 200 to 300 holes per square inch. These holes follow the softer wooden rings and are .05 to .025 inch in diameter, seldom over 3/4 inch long. As a result, the wood becomes a mass of thin walls between burrows which break away exposing new areas to attack. In this manner the pile is slowly reduced in diameter (Fig. 34). In soft woods like pine and spruce, the diameter may be reduced as much as two inches per year. The chief area of attack is between the low water level and the mud line with occasional activity up to the high water level. The limnora does not seem to be affected by small environmental changes and may be found in salt or brackish water of any temperature, polluted or clean.

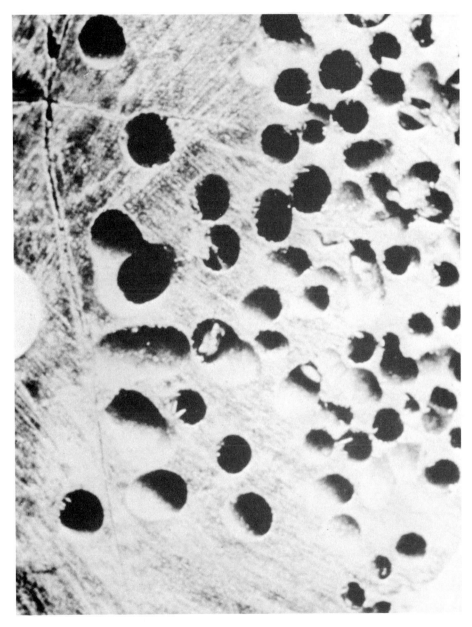

FIG. 32: Cross-cut Timber Member Showing Extensive Damage

FIG. 33: Insect Attack Resulting in Complete Failure

FIG. 34: Reduction at Water Line Due to Insect Attack

53

Complete protection from borer attack is essential. Metal armoring is falling into disuse as is concrete casing. The very best method may prove to be jacketing creosoted piles. The most practical method involves heavy treatment with high quality, coal-tar creosote by the full cell method to the point of near saturation. Although shipworms will generally not attack a creosoted member, they may attack any area which has been damaged and any untreated area. Limnora attack the creosoted timber directly but at a decelerated rate. Other types of wood preservatives in used include a plastic outer wrap which is successful as long as the plastic is not damaged and cyanide treatment which is messy to apply, leaches out with time, and may, therefore, cause damage to the environment.

Fire damage to wood is a real problem on bridge and wharf structures. Treatment of wood with preservatives may protect the wood from fungi and insect attack but generally results in making the wood more susceptible to fire. Damage due to fire is readily apparent due to the charred appearance and burnt odor of fire damaged wood.

Chemical damage to wood is generally very difficult to determine because it often resembles damage done by other factors. Fungi attack employs a chemical reaction between the enzymes particular to each fungus and the wood fiber. Fire damage involves the oxidation of the wood fiber which is again a chemical process. It is, therefore, important to report any apparent damage to the timber as soon as possible.

Impact damage may occur in the event of a high energy collision between a vessel and a fendering system. Because wood has good impact characteristics, it may show only limited signs of external damage and must, therefore, be inspected carefully after an accident or suspected accident. The timbers may have a shattered appearance as opposed to the sagging appearance caused by overload. Compression failure will resemble wrinkled skin while tension failure looks as if the fibers have been pulled apart.

Once provisions have been taken to insure that failure will not occur as a result of the factors discussed above, a wood must be selected on the basis of its material properties. The timber must have a high compressive strength perpendicular to the grain to resist crushing, a high degree of fiber hardness to resist rubbing action but not to the point that brittleness and checking result, and a relatively high bending strength. In general, Douglas fir, Southern pine, White oak, and Red oak are commonly used in the United States. These woods must be protected in the manners discussed earlier, but have good workability, are generally readily available and inexpensive. Douglas fir is the only structural timber from which high quality piles of 100 foot lengths can be obtained.

Various hardwoods have been used in fendering and include several exotic species with natural borer resistance. Greenheart is the most common of these woods and is available in various thicknesses and lengths. It is expensive. Tropical timbers tend to be much harder than the soft woods and cannot be worked with ordinary tools intended for soft woods. Metal working tools are necessary as are pre-drilled holes. This extreme hardness may result in splintering of the timber fiber. Greenheart is heavier than water resulting in additional construction difficulties, but has a service life more than three times that of oak members.

The choice of Greenheart or the softer woods must be based on economic, construction, preservation, and service life considerations. There is no single solution to this debate. The choice must be left with the designer and the engineer who have knowledge of the peculiar problems of their design site.

STEEL

Steel is a material which has a life long high compressive and tensile strength, but its use in marine structures is limited.

Man's concern with the corrosion characteristics of metals used in maritime works dates to the fourth and fifth centuries B.C. It was observed that while most metals used in ships deteriorated in some way when exposed to sea water, bronze showed little or no corrosion. Since the use of bronze in marine installations is not feasible now or in the near future, methods must be found for protecting the metal most commonly used in these structures, steel, from salt sprays, tidal action, and other environmental factors.

Corrosion refers to the conversion of metals into compound forms as a result of some natural mechanism. The most common types of corrosion are galvanic, differential environment or concentration cells, stray current, bacteriologic, stress, fretting, impingement, and chemical corrosions. Galvanic corrosion requires an electrolyte, sea water, and two dissimilar metals coupled together, one of which will corrode. The corroding area is referred to as the anode, while its counterpart is the cathode. The process involves the flow of current from the anode, which will corrode, to the cathode, which is protected. Certain metals have been used as anodes to steel: magnesium, zinc, and aluminum to name a few. The order in which joined metals in an electrolyte will corrode is determined by the voltaic potential of the metal. Magnesium will corrode to all other metals. Zinc will corrode to all other metals except magnesium. Aluminum will corrode to all other metals except magnesium and zinc, etc. Corrosion producing galvanic cells may occur in the same metal. Steel is anodic to mill scale found on hot rolled steel products such as steel piling. Bright surfaces are anodic to matte surfaces. Stressed surfaces are anodic to normal surfaces. In addition, surface imperfections and non-homogeneous metal may cause galvanic cells.

Differential environment corrosion occurs when a metal such as steel is in contact with an electrolyte which is non-homogeneous from point to point or varies in concentration along the surface of the metal. Some of the major environmental conditions which may cause corrosion cells to develop on steel pilings include varying salt conditions in the soil in contact with the pile, varying moisture conditions in the soil in contact with the pile, different oxygen concentrations in various horizons of the soil or water in contact with the pile, and foreign matter embedded in the soil in contact with the pile such as cinders or pieces of metal. Within the area of tidal fluctuations, the oxygen content varies considerably between the high and low tide levels and the low tide and ground levels. Current flows from the low oxygen to the high oxygen content area resulting in corrosion in the low oxygen or anodic area. Protruding elements, corners, and weld points tend to corrode faster than the surrounding surface metal because the interruption of flow results in development of an oxygen concentration cell. In those areas where fresh water meets sea water, corrosion of steel piles are likely. The higher oxygen content of the fresh water causes current to flow from the salt water area to the fresh water area producing anodic conditions in the salt water and subsequent corrosion of steel pilings in the salt water area.

Stray electrical current corrosion is an electrochemical process similar to galvanic and differential environment corrosion. It is the result of stray electric currents from electric railways, railway signal signs, cathodic protection systems for pipelines or foundation pilings, DC industrial generators, or DC welding equipment. Since metals are more conductive than water or soil, stray currents tend to collect in the metal, following the path of least resistance. Corrosion occurs when the current leaves the metal producing stray current electrolysis at the point where the current exits the metal. While stray alternating currents are generally not harmful to underground or underwater structures, the current can be rectified by passing through or off metal structures. The partial DC component is then capable of producing current electrolysis, as discussed earlier. The sources of alternating current include central power stations and large substations.

Bacteriological corrosion is caused by two types of micro-organisms, anaerobic and aerobic. Of the two types, the anaerobic is the most responsible for the corrosion of steel. This corrosion results from the creation of concentration cells on the surface of the metal, the creation of a corrosive environment through water and decomposition, and depolarization of cathodic or anodic areas. Corrosion due to this particular type of organisms may be found in bogs, heavy clay, swamps, stagnant waters, and contaminated waters, particularly when contaminated with the waste from paper-making operations. The external corrosion products are black underlain with a thin film of white substance.

Stress corrosion requires high stress in most instances although this may vary with the environment. The stress may be residual stress from cold working, or may be due to applied loads or welding operations. Tensile forces expose metal at the grain boundaries resulting in small intergranular cracks. As the corrosion continues along the grain boundaries, the combination of stress and continued corrosion results in transgranular cracking.

Fretting corrosion occurs on closely fitted metal parts which are under vibration such as a sleeve on a shaft, shrunk metal fittings, or forced metal fittings. The contact surfaces may become pitted and a deposit, usually red in color, may form at the interface. The corrosion damage affects operating tolerances and increases the possibility of failure.

Impingement corrosion affects brass and hard drawn copper and is, therefore, of little concern when dealing with maritime installations.

Chemical corrosion results from the direct attack on steel by acids or diluted acids. Chlorine, carbon dioxide, and sulfur oxide are atmospheric contaminants which may cause the formation of acid films on metal surfaces resulting in rapid corrosion. Chlorine is more commonly found in the marine environment while carbon dioxide and sulfur oxide are common industrial contaminants.

The rate of all of these corrosion reactions is directly affected by the concentration of dissolved oxygen in the water, the concentration of corrosion products in the water, and environmental factors which may affect the condition of the electrolyte such as temperature, salinity, etc. The most rapid corrosion occurs in the tidal fluctuation zone where dissolved oxygen is plentiful and the moving water body prevents saturation of the water surrounding the metal with corrosion products or the deposition of corrosion products on the surface of the metal.

In many instances the failure of a steel structure in sea water could be prevented by thorough design investigations and considerations. These considerations should include the selection of materials to suit the environment, the minimum galvanic potential between two metals, the relative areas of dissimilar metals which are to be joined, the resistivity of the soil and water surrounding the metals, the possibility of applying insulating coatings, and the prevention of water accumulations where feasible. These design criterion may reveal the need for external coatings applied to the surface of the steel members.

Coatings may take many forms: metallic coverings and claddings, organic coatings and wrappings, synthetic organic materials, and concrete casings. Metallic coverings and claddings protect the base metal in two ways depending upon the position of the applied metal in the electromotive force series. If the applied metal is more corrosion resistant than the base metal, the

59

applied coating is referred to as sacrificial and will corrode to provide
protection for the base metal. Sacrificial coatings consist of zinc or aluminum. Organic coatings and wrappings resist moisture absorption and isolate
the metal structure over long periods of time from the forces of electricity.
Some of the common organic materials used as coatings and wrappings include
asphalt, coal tar, paints, plastic wrappings, and greases. Of interest in
recent years has been the effect of oil spills on the corrosion resistance
on metal structures. Oil films which develop on the structures provide ex-
cellent protection from corrosion by isolating the structure from air, and
water. This method of protection tends to be too costly to be used as a
systematic corrosion preventative. Synthetic organic materials have been
widely used for electrical insulation and isolation. Plastics are also used
in the protection of pipelines from underground corrosion and at the joining
points of dissimilar metals to prevent current flow. Concrete casings are
the last area of external applied protection. The characteristics of the
concrete necessary to insure adequate protection is discussed in the section
of this chapter dealing with concrete corrosion. The concrete may take the
form of shotcret or gunite but must have a minimum depth of cover of three
inches. External coatings tend to deteriorate with age resulting in checking,
flaking, scaling, chalking, washing, blistering, peeling, cracking, or spalling.
They may, however, be the most economically feasible method of protection for
a particular structure.

Cathodic protection refers to the protection of steel or other metals
through electrolysis by stray electric currents. There are two methods of
cathodic protection, sacrificial and impressed current; each uses a base metal
which is corroded instead of the steel in the structure. In the sacrificial
system the metal to be protected is wired to a more negative metal and both
are placed in an electrolyte, either soil or water. Current flows from the

anode to the cathode, from the more negative metal to the less negative metal resulting in the corrosion of the anode and protection of the cathode. Magnesium and zinc are commonly used in the sacrificial system. The impressed current system requires an external power source and anodes consisting of any conducting material suitable for the purpose. Current flows from the anode connection to the power supply to the cathode which is a relative ground. The variable power source provides a wide range of voltage and current possibilities although the entire system is subject to failure due to power failures. The required maintenance and inspection of the impressed system is far greater than that required by the sacrificial system. The sacrificial system is relatively simple to install and maintain. Additional anodes can be added without revamping the design for the entire system. Both systems are effective in protecting those portions of the structure which are constantly immersed in water or imbedded in the ground. Neither system is effective in the splash zone, the area of greatest corrosion, or above water or ground level.

To insure the greatest protection in the splash zone, heavy concrete casings and/or corrosion resistant metal shielding are recommended. Further research into effective corrosion prevention methods should pay special attention to splash zone protection and feasibility of application in areas where little maintenance is likely to be performed.

Good maintenance of the structure can make or break any protection system yet devised. The controlling variables in atmospheric corrosion are the length of time that the metal is wet; the amount of foreign matter in the atmosphere, particularly chlorine, sulfur, and carbon dioxide which form acid films; and the composition of the metal. Proper maintenance can do little as far as controlling the composition of the metal or the foreign matter in the atmosphere, but it can affect the length of time that the structural members are wet. Structural members can be kept free of dirt and debris which retain rain and wash water as well as the deicing salts which may be present, and may even absorb moisure and corrosive particles from the atmosphere. The water and salts absorbed by the dirt and debris is held in contact with the metal for long periods of time resulting in accelerated corrosion.

Deterioration of a steel structure may result from processes other than corrosion although for fender systems corrosion is of primary significance. Fatigue cracking may occur at connections and points on the structure where a discontinuity or restraint is introduced; at loose connections or members which could force a member to carry unequal or excessive stress; at damaged members, regardless of magnitude, which are misaligned, bent or torn; at sites where corrosion could reduce load carrying capacity through decrease in the member section making it less resistant to both repetitive and static stress conditions; at weld points where crack initiation might begin; at sites of part repair and areas of excessive vibration or unusual twisting; and at places where structural details are known to have exhibited fatigue problems. The effect of high temperatures on steel strength becomes of importance when considering the possibility of fire on a dock or pier structure. During regular operating conditions, the temperatures remain within the realm of elastic design and pose no unusual design problems. Careful inspection of the structure can generally reveal areas of fatigue cracking or plastic failure due to high termperatures.

CONCRETE

The use of concrete in marine structures has increased in recent years as the size of berthing vessels has increased. Concrete has high compressive strength but low shear and tensile strength. It is porous, extensible, and fire resistant, but can be damaged by intense heat. Under ordinary loading conditions, concrete is elastic though it will creep under sustained, heavy loads. The shear and tensile strength of concrete can be increased by the addition of reinforcing bars or by prestressing with steel wires which are under tension. This factor allows much design flexibility yet deterioration of concrete in the marine environment can pose significant problems.

The factors which cause deterioration may be grouped under the following areas: poor design details, construction deficiencies and operations, temperature variations, chemical attack, reactive aggregates and high alkali cement, moisture absorption, wear or abrasion, shrinkage and flexure forces, collision damage, scouring, shock waves, overstress, fire damage, foundation movement, and corrosion of reinforcing bars and prestressing wires. The visual symptoms of deterioration consist of cracking, scaling, spalling, rust stains, surface disintegration, efflorescence and exudation, and vehicular damage. Each shall be examined with the form of deterioration most responsible for that form of disintegration.

Poor design details which can cause concrete to crack include insufficient drainage. Scuppers may not be provided or improperly provided with downspouts to keep the water discharge away from concrete surfaces including caps and decks. The scuppers may be too small and, as a result, easily clogged. They should be provided at all low spots. Weep holes, if provided, may be too small, too few, or discharging over another concrete surface. When not enough space is provided at an expansion joint, spalling may result (Fig. 35). Insufficient cover over rebars may cause corrosion of the rebars, or, in the case of prestressed concrete, the prestressing wires. Of all of these considerations, the need for sufficient expansion space cannot be overstated.

Construction deficiencies may result in concrete deterioration regardless of the care taken in the design procedure. Soft spots in the subgrade of a foundation will cause settlement resulting in cracking (Fig. 36). If formworks

FIG. 35: Spalling at Expansion Joint

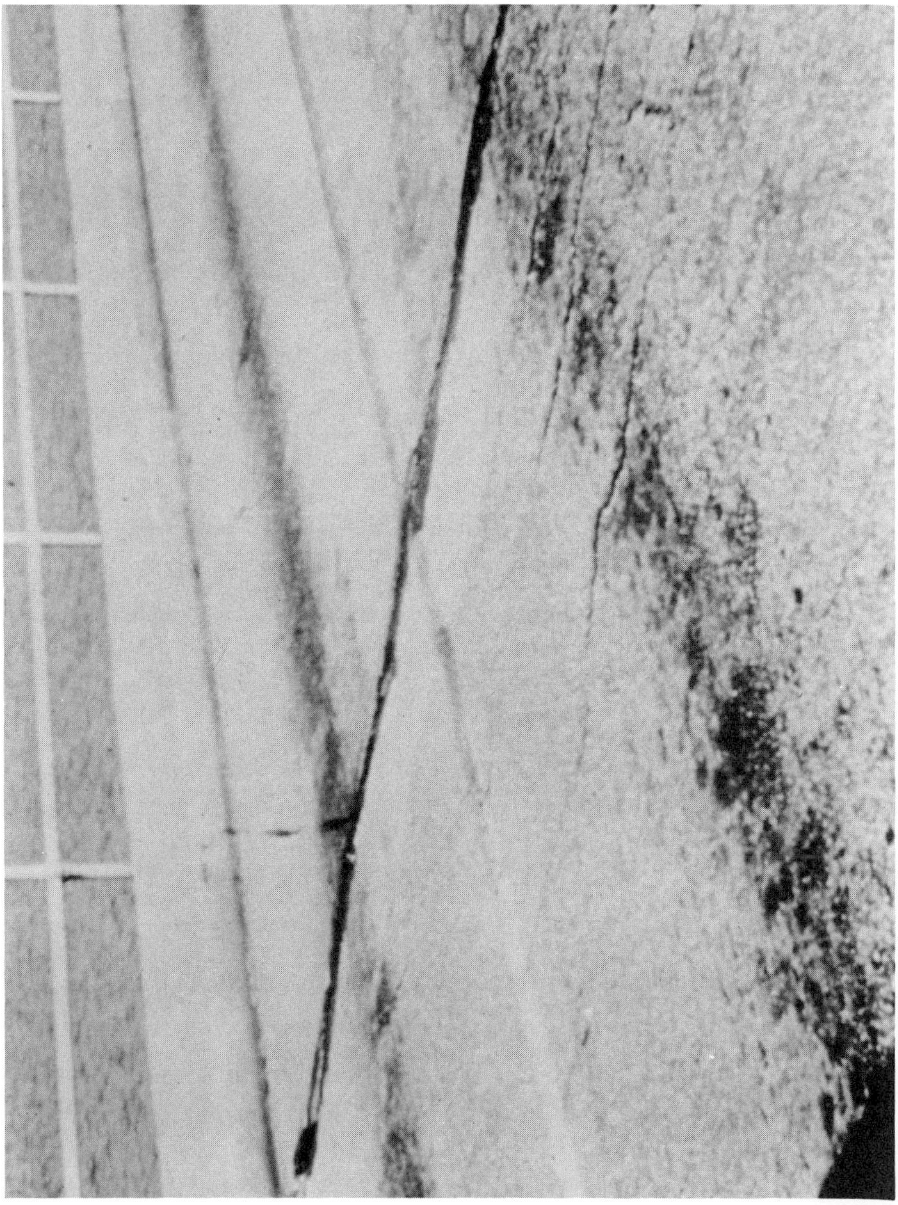

FIG. 36: Cracking Due to Foundation Settlement

are removed between the time the concrete begins to harden and the specified time for formwork removal, cracks will probably occur. These cracks are especially dangerous since they may occur internally with no external manifestations. Water can collect in these cracks and cause spalling due to freezing and thawing or cause corrosion of any internal steel which may eventually lead to cracking and spalling. If sufficient spacing does not exist between rebars in reinforced sections, voids may develop if the mix is not properly vibrated. These voids collect water with end results similar to those discussed earlier for cracks which collect water. Excess vibration may cause segregation of the concrete mix, resulting in a weaker concrete than specified. The inclusion of clay or soft shale particles in the concrete mix will cause small holes to appear in the surface of the concrete as these particles dissolve. These tiny holes increase the porosity of the concrete and, as before, lead to cracking and spalling and possible corrosion of the internal steel (Fig. 37).

Freezing and thawing (temperature variation) is a common cause of concrete deterioration. Porous concrete absorbs water which expands as a result of freezing. The internal expansive pressures developed in this manner often produce cracking (Fig. 38), spalling (Fig. 39), or scaling. Scaling appears as a chalky matrix typically white. Temperature fluctuations may further affect concrete integrity if the coefficient of thermal expansion differs significantly from that of the mortar. Aggregates with lower coefficients may cause high tensile stresses resulting in cracking and spalling. Further problems arise if the concrete section is restrained from expansion or contraction. The internal forces set up under these circumstances are sufficient to result in cracking and spalling and eventual failure of the member.

Chemical attack of concrete may come from two sources. The use of salt or chemical deicing agents contributes to weathering through recrystalization, and salt may increase the water retention. The results are similar to the

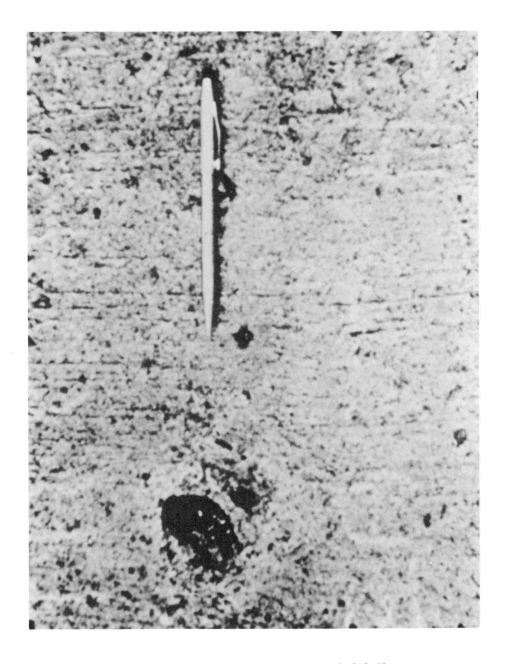

FIG. 37: Surface Pitting Due to Included Clay

FIG. 38: Freeze-Thaw Cracking

FIG. 39: Surface Deterioration

effects of freezing and thawing: cracking and spalling. Further chemcial
attack may come from chemicals in the soil or water surrounding the concrete
members. Ammonium and magnesium ions react with the calcium in the cement
paste. Sodium, magnesium and calcium sulfates react with the tricalcium
aluminate in the cement paste. Acids will attack the cement paste by chem-
ically transforming the composition of the paste. The results of general
chemical attack appear as surface disintegration in the form of scaling and
spalling, random cracking in unrestrained members, swelling and parallel
cracks in compression members, and protruding aggregate resulting from the
removal of cement paste.

Reactive aggregates and high alkali cements cause swelling, map cracking
(Fig. 40), and popouts (Fig. 41).

Moisture absorption will cause concrete to swell. Concrete cylinders,
thirteen feet in diameter, have grown as much as six inches in a marine environment. If restrained, the restraining material will break apart or the
concrete will crack and spall. In addition, soft water will tend to leach
out the lime in the cement and leave a powdery residue.

Wear and abrasion cause the surface of concrete to disintegrate over
extended periods. Wind and water-driven particles, particularly sand, play
a significant role in abrasion near the mud line and above the high water
line. Piles may be damaged due to the rubbing action of vessels resulting
in scaling, revelling, and cracking at joints, and scarring.

Shrinking and flexure forces, common design considerations in concrete
piles, may set up tensile stresses exceeding the capacity of the section.
The end result is cracking. Setting shrinkage generally causes shallow surface cracking (Fig. 42). Drying shrinkage may take place over extended
period of time producing tensile forces and eventually cracking. The time
frame may amount to several years.

FIG. 40: Map Cracking

71

FIG. 41: Popouts Due to Reactive Aggregates

FIG. 42: Shallow Surface Cracking

Collision damage is generally considered to be an accidental occurence. Any concrete structure which can be reached by a moving vehicle has suffered collision damage at some time. In most instances the vehicle is more severely damaged than the structure. After accidental collisions the structure should be inspected for possible damage.

Concrete surfaces in surf zones are scarred by sand and silt (Fig. 43). Ice flows in rivers and bays are also responsible for considerable damage to concrete pilings and piers (Fig. 44). Most damage of this type occurs between the low and high water marks.

A shock wave can damage concrete due to the varying transmission rates through the aggregate, the paste and the reinforcing steel. The shock waves then become partially additive setting up conditions leading to cracking and spalling of the concrete mass. Concrete piles are vulnerable to damage from shock waves while they are being driven.

Overstressed concrete may exhibit longitudinal and lateral cracking in a deck. Over a bearing point there may be diagonal cracking at the end of a simple beam (Fig. 45), vertical cracking running from the bottom at the center of a simple beam to the neutral axis, and vertical cracking from the top of the beam extending downward to the neutral axis for a beam which is continuous over the bearing area (Fig. 46).

Fire damage results from the extreme temperatures of a large fire. Temperatures above 300° C. will cause a weakening in the cement paste and lead to cracking and spalling (Fig. 47).

Foundation movements will cause serious cracking in concrete structures if they generate a sizeable tensile stress in the concrete piers or abutments Vertical cracking predominates.

Corrosion of the internal reinforcing steel causes tremendous expansion pressures in the concrete. Fully corroded steel occupies seven times the

FIG. 43: Sand and Silt Scarring

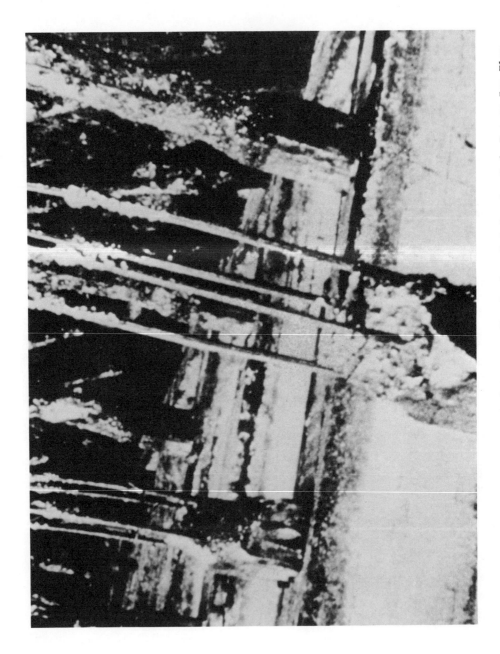

FIG. 44: Total Removal of Concrete Matrix from Reinforcing Rods Due to Ice Flow

FIG. 45: Diagonal Cracking

77

FIG. 46: Concrete Member in Models Lab

FIG. 47: Cracking and Spalling Due to Fire

79

volume of uncorroded steel, therefore, cracking and spalling result from the volume increase (Fig. 48). Corrosion of the steel may also result if water finds its way into the concrete member.

Corrosion of the prestressing wire combined with the high tensile stress required for prestressing may result in failure of the member due to stress-corrosion. Failure of a significant number of prestressing wires in this manner will result in loss of tensile strength in the member and could lead to failure under heavy loading conditions. Further deterioration of the tensile strength of a member may result from creep of the prestressing steel, shrinkage of the concrete causing a relaxation in the prestressing wires and creep of the concrete shortening the overall length of the member and thereby relaxing the prestressing wires. Other causes for a loss of prestress include elastic deformation of the concrete, drawing in of the anchorages, and friction loss in the post-tensioning operations. The combined loss of prestress due to such causes can amount to as much as 25 percent to 35 percent of the member's design strength. The result of loss of prestress is cracking particularly near the anchorages and on the compression face. Unlike cracks in high tension areas of reinforced concrete members, the appearance of cracks in a prestressed member may have serious effects on its structural integrity. A prestressed member is usually under high compressive stresses, consequently cracking should not be expected.

With proper design mixtures, it is possible to produce concrete with service life of fifty years in a marine environment. The concrete should be dense with an impervious surface and relatively nonabsorbent characteristics. All internal steel should be covered to a depth of at least three inches of good concrete. The cement content should range between a minimum of 6-1/2 sacks per cubic yard and a maximum of 7-1/2 per cubic yard. The aggregates should be graded for maximum density and nonreactive. If nonreactive aggregates

FIG. 48: Piles Cracking

are not available, then pozzolans must be added to the aggregates or a low-alkali cement must be used. The water content should be such as to provide a workable, plastic mix with the lowest water-cement ratio possible, not to exceed six gallons per sack of cement. Type V cement with five percent maximum tricalcium aluminate should be used although Type II may be substituted under tight economic conditions. Three to six percent entrained air is usually specified for aggregate size ranges from 1-1/2 to 3 inches. Thorough curing by the best means possible and adequate vibration is necessary to insure the concrete reaches its design strength. Particular care should be taken when handling precast members. Support is necessary along the entire length of reinforced members due to low bending strength. Prestressed members are less susceptible to handling stresses than ordinary reinforced members but should be treated with some degree of reasonable care.

Other forms of concrete used in the marine environment include asphalt impregnated concrete. The outer shell of the concrete is impregnated to a depth of about one inch to three inches or more. The method of application resembles that for creosoting wood. The concrete is placed in a bath of molten asphalt admitted under high vacuum after being pre-dried at temperatures up to 250° F. The member is then subjected to air pressures up to 150 psi. The asphalt is then removed and the concrete is exposed for several hours to pressures of 100 psi. It is then removed to a cooling chamber. During the entire process the temperature gradient is closely watched. Asphalt impregnated beams and cylinders have been found to be dry after storage in the ocean at an elevation of -35 feet for 35 years. This type of pile has been used in the Los Angeles harbor since 1925 with no symptoms of deterioration to date.

Shotcrete, also known as gunite or spraycrete, is another form of concreting which has been used with success in the marine environment. It is usually proportioned by volume in the ratio of one part cement to four parts well graded sand although richer mixtures have been used. The richer mixtures have an increased tendency to cracking as a result of temperature variations and alternate wetting and drying. Twenty-eight day compressive strengths of 4000 psi are possible with the leaner mixtures, well within the requirements of most design structures. The use of shotcrete instead of conventionally placed concrete can result in savings of material and time, as well as providing a more uniform and dependable member if the field control of the conventional formwork pouring is limited or inadequate. It has very good bending characteristics with older concrete and is, therefore, used for patching and plugging of existing structures. The critical parameter in the success of a shotcreted member is the skill of the craftsman applying the shotcrete. The angle of the nozzle, the rate of application, the amount of pressure on the line, the distance of the nozzle from the member, all are critical to the success of the structure. The final design decision to use conventionally placed concrete or shotcrete can be made on the basis of economy since the deterioration rates are quite similar. One disadvantage of the shotcrete method is that it cannot be used to produce a massive member.

CHAPTER IV

DESIGN PARAMETERS

The function of bridge fendering systems is to protect bridge elements against damage from waterborne traffic. There are many factors to be considered in the design of fendering systems including the size, contours, speed, and direction of approach of the ships using the facility, the wind and tidal current conditions expected during the ship's maneuvers and while tied up to the berth, and the rigidity and energy absorbing characteristics of the fendering system and ship. The final design selected for the fender system will generally evolve after making arbitrary limitations to the values of some of these factors and after reviewing the relative costs of initial construction of the fendering system versus the cost of fender maintenance and ship repair. It will be necessary to decide upon the most severe docking or approach conditions to protect agginst and design accordingly; hence, any situation which imposes conditions more critical than the established maximum would be considered in the realm of accidents and probably result in damage to the dock, fendering system, or ship.

The kinetic impact energy may be simply expressed as;

$$E = \frac{1}{2} mv^2 \tag{1}$$

where E is the energy of the system of mass, m, moving at a velocity, v. Each of these variables involves a complicated set of interrelationships bearing directly on the design of an adequate fendering system.

The total energy, E, may be expressed as the sum of many component energies: E_{f-d}, the energy absorbed by the fender and dock structure, E_s,

84

the energy absorbed by the ship, and E_{Lost}, the energy lost due to friction
with the harbor bottom, friction in the fender-dock system, etc. While the
energy lost in unexpected forms may cause problems requiring dredging and
other physical modifications, it cannot be accurately evaluated due to its
random nature.

Energy absorption in the ship may occur in two basic ways: by deformation of the hull, E_d, or by rebound of the ship, E_m. For the purposes of
simplification, deformation will be assumed to occur in the elastic regime.
The energy of the ship can, therefore, be expressed as;

$$E_s = E_d + E_m \qquad\qquad (2)$$

E_m is the function of the radius of gyration of the vessel, the vessel geometry, sea conditions, and the velocity and deceleration of the vessel immediately prior to impact. Accurate analytical determination of E_m is impossi-
ble. It is, therefore, necessary to evaluate the importance of this energy
to the system, and, if possible, justify its elimination from consideration.

The transfer of energy to the fender at the moment of impact results in
deflection of the fender. At some small increment of time immediately after
this, some energy is transferred back to the vessel in the form of the energy
of rebound. The rebound energy is a function of the stiffness of the fender
which determines the characteristics of the collision as well as the deflection of the fender under various loading conditions. The result of this rebound is translation and rotation of the vessel in three dimensions causing
pitch, yaw, and roll. Maximum rebound energy requires the collision to be
completely elastic resulting in no energy absorption by the fender. If this
were the case, severe damage might be incurred by the vessel due to the limited stress-strain characteristics of any finite material element. Minimum
rebound energy would occur for a totally inelastic collision where all energy

85

is absorbed by the fendering system. A system designed for minimum energy case with a finite deflection range would tend to be stiff resulting in possible damage to the vessel. Between these two extremes lie the actual case. Conservative design measures call for no rebound by the ship or an assumption of the ratio of the rebound energy to the hull deformation energy. If E_{Lost} is assumed to be negligible, the total energy equation reduces to;

$$E = E_d + E_{f-d} \tag{3}$$

The impact energy is a time related function of impulse. The effects of impact on a ship hull may be divided into three time increments and examined separately. During the period of localized effect, between 0 and 1/200 second, stress and deformation characteristics are determined by the dynamic elasticity of plate panels, frames, stringers, girdles, etc. The resulting three dimensional equations can be solved only for very simplified, idealized structures with results of questionable value for the actual physical situation. During the transition period, from 1/200 to 1/20 second, a stress-deformation wave travels through the ship. After 1/20 second a vibratory motion is established. Damping of these oscillations results from internal friction in the steel hull, fluid friction in the boundary layer adjacent to the ship, wave or ripple formation on the ocean surface, and friction and slip in the lab and butt-riveted joints of the hull. These factors have been evaluated both experimentally and analytically assuming a simple beam structure. Experimental results yield decremental damping time functions three to ten times those obtained by simple beam assumptions.

Research into the area of the impact response is needed to shed light on the complex behavior of the ship. Several expressions which could be used to develop scale models for research include equations for velocity of the shear-bending wave in the hull, V_{s-b}, the frequency, w, and the impact load (F(t):

86

$$V_{s-b} = \sqrt{G_g/\rho_H} \tag{4}$$

where V_{s-b} is the velocity of the shear-bending wave in the hull, G is the shear modulus of elasticity, g is the acceleration of gravity, ρ_H is the density of the hull material;

$$\omega = \sqrt{EI_g/\rho_H A_H L^4} \tag{5}$$

where ω is the frequency, E is the modulus of elasticity, I is the moment of inertia of a cross section about the neutral axis, g is the acceleration of gravity, ρ_H is the density of the hull material, A_H is the cross-sectional area of the hull, L is the length of the hull;

$$P(t) = \rho_{f-d} V^3 t L_s \tag{6}$$

where P(t) is the impact load, ρ_{f-d} is the equivalent fender-dock material density, V is the velocity of impact, t is the time measured as t=0 at impact, L_s is the length of the hull in contact with the fender.

The energy absorbed by the fender is a function of the stiffness of the fender and the impact energy. If the stiffness of the fender is expressed by the value k and the deflection by d, the energy absorbed by the fender, E_f, is;

$$E_f = kd^2 \tag{7}$$

Complications arise from the fact that k is generally not a constant but a function of deflection. Appendix I contains the graphs of k vs. d curves. The Appendix also contains energy tables listing energy levels as a function of the percent deflection based on the initial size of the fendering device. The Appendix is grouped by companies for easy reference. The data used to generate the curves and tables was taken from design aids published by the manufacturers.

The evaluation of the energy absorbed by the dock or pier structure

87

introduces further complexities to the system. The structural components have varying stiffnesses, deflection limits, geometric shapes, external operating parameters, etc. As the sophistication of the structure increases, so do the number and complexity of the energy equations. Various simplifying assumptions have been proposed. The most popular to date consists of replacing the structural members with springs and rigid plates with non-linear functions incorporated in the spring constant expressions. This method lends itself to computerization and allows complete analysis of the entire system once a spring-rigid plate model is developed and spring constants are determined. In this manner the energy absorbed by the structure can be expressed in terms of various stiffnesses and deflections dependent upon the impact energy of the vessel and the fraction of the energy which might be absorbed by the ship. In this analysis also, the difficulty of evaluating the proportionality of the energy absorption of the ship and structure again becomes evident.

The need to determine this proportionality has led to the use of many equations of the form;

$$E_f = C \times E \qquad (8)$$

where E_f is the energy abosrbed by the fender, C is some constant which reflects the amount of energy which is not absorbed by the ship or pier structure and must, therefore, be absorbed by the fender, and E is the kinetic energy of impact. The constant, C, is a function of the berthing point of the ship, the geometry of the ship, the hull stiffness, the type of dock structure, the sea exposure conditions, and the hydrodynamic mass, or the added mass of water set in motion by the ship.

The components of the energy constant, C, can be grouped into various coefficients; an eccentricity, C_E, a stiffness coefficient, C_S, a configuration coefficient, C_C, and a hydrodynamic coefficient, C_H. C is as follows:

$$C = C_E \; C_S \; C_C \; C_H \qquad (9)$$

The eccentricity coefficient, C_E, adjusts the impact vector, which is generally not along the velocity vector, to that portion which is normal to the pier. The rest of the energy is utilized as rotational energy. If the assumption is made that the geometric center of the vessel is also the center of gravity, an expression can be written which relates the radius of gyration of the ship about the vertical axis through the center of gravity, K, with the distance along a line joining the center of gravity and the point of impact, r, and the angle between r and the velocity vector, ϕ;

$$C_E = \frac{K^2 + r^2 \cos^2\phi}{K^2 + r^2} \qquad (10)$$

Figure 49 is a schematic diagram of this non-sliding contact of the ship and the fender, r is the line joining the center of mass with the impact point; V is the velocity vector; ϕ is the angle between r and V and C.G. labels the center of geometry. This may differ from the center of gravity due to the loading condition of the ship. An exact calculation of the center of gravity would require detailed knowledge of the loading conditions and cargo of the shipping traffic, as well as extensive knowledge of the architecture of the ship. To determine the center of geometry a plan view of a representative ship is required. From this view the center of gravity can be located and the analytical solution of equation 10 is possible.

An approximate solution can be obtained if the velocity vector, V, is assumed to act perpendicular to the line joining the center of gravity and the impact point: $\phi = 90^\circ$. Equation 10 then becomes:

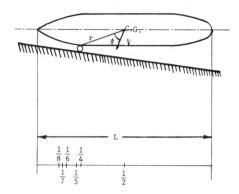

FIG. 49: Schematic Diagram of Ship Contact

$$C_E = \frac{K^2}{K^2 + r^2} \qquad (11)$$

or

$$C_E = \frac{1}{1 + (r/K)^2} \qquad (12)$$

Figure 50 is generated by solving equation 12 for various berthing points measured as a function of the ship length, L, as illustrated in Figure 49. Commonly used values are: for impact at a third point (1/3 the length of the ship), $C_E = 0.5$, and for impact at a quarter point (1/4 the length of the ship), $C_E = 0.7$, from Figure 50.

The configuration coefficient, C_C, is a function of the supporting structure and is generally assumed equal to 1.0 for an open pier, 0.9 for a semi-closed pier and 0.8 for a closed pier. The open and closed configurations are illustrated in Figure 51.

The stiffness coefficient, C_S, is generally assumed to be 0.9 as a conservative estimate of the percentage of energy which the fender must absorb based solely on stiffness relationships.

The hydrodynamic mass concept attempts to account for the added forces, pressures, and impulses that arise from the motion of a ship through some body of water. This added mass of water is considered to be added to the mass of the ship and said to govern the motion of the ship. This type

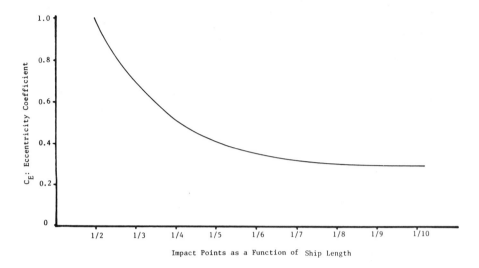

Impact Points as a Function of Ship Length

FIG. 50: Eccentricity Coefficient vs. Impact Points

of approach has been used to account for those factors not covered by the other parameters. The effect of consideration of the added mass depends on many factors: waves, winds, nature of impact, character of the structure, flow of water around the ship and structure, etc. A demonstration of the variability of this factor requires only a brief look at the berthing man-euvers of a ship approaching the dock at two speeds. In the first case, the approach is rapid and braking of the forward motion occurs suddenly. The water is forced to flow around the vessel resulting in some forward motion and impact force due to the mass of water. In the second case, the velocity is low and braking occurs gradually. The mass of water does not split to flow around the vessel and the effects of the moving water body are almost negligible.

Past design techniques have described this added water mass, M_W, as a cylinder whose diameter is the draft of the ship, D, and whose length is the length of the ship, L.

$$M_W = \pi/4 \ D^2 L \rho_W \tag{13}$$

91

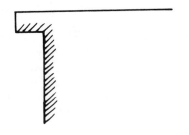 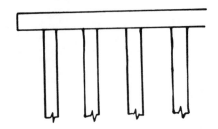

(a) Closed dock structure or
    pier structure

(b) Open dock structure or
    pier structure

FIG. 51: Open and Closed Dock and Pier Structures for
the Determination of the Configuration Coefficient

where ρ_W is the density of the water and may vary with geographic

location.   The hydrodynamic coefficient in this analysis is the ratio of

the added water mass to the displaced mass of the ship, M_S:

$$C_H = M_W/M_S \qquad (14)$$

The ratio of M_W to M_S, or C_H, has been correlated to the ratio of the beam

of the ship, B, to the draft of the ship, D, and is shown in Figure 52a.

Other design methods involve the use of empirical equations relating

C_H to B/D.   The most commonly used is

$$C_H = 1 + 2 \; D/B \qquad (15)$$

The second term in this expression, 2D/B, is shown in Figure 52b and is

labelled to 2 D/B.   Adding 1.0 to the value read from this graph gives the

C_H value for design.   Another suggested equation is

$$C_H = 0.3 + 1.8 \; D/B \qquad (16)$$

This equation is shown in Figure 52b and is labeled 0.3 + 1.8 D/B.

The value read directly from this graph is the C_H design value.

Figures 52a and 52b allow quick visual determination of C_H values when

these values are to be incorporated in design considerations.   All of these

approaches for the determination of C_H are approximations and differ from

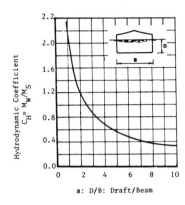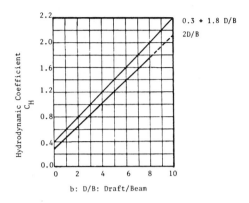

a: D/B: Draft/Beam   b: D/B: Draft/Beam

FIG. 52: Hydrodynamic Coefficient vs. Draft to Beam Ratios

measurements using scale models and actual berthing maneuvers. Model studies and berthing maneuvers yield coefficient values ranging from 1.3 to 3.0.

Values presented by various companies in their design aids range from 1.5 to 1.9 as compared with the model-berthing values and the approximate values from equations 14, 15 or 16. Further research into this area is indicated although the variability of the impact velocity and slowing rate may be controlled by human judgment and experience.

The velocity of impact of vessels has received much study in the past few years. It is affected by currents, wind acting on the freeboard area, boundary layer conditions, and pilot experience. It appears that while smooth curves may be forced through data points relating velocity to other factors, the single largest factor governing the impact velocity is the nautical judgment of the pilot during berthing or passing maneuvers. While some authors have suggested an inverse relationship between size and speed, practical experience has shown that reductions in the design velocity based on this assumption may result in overloading the designed structure during the normal range of operations. The following values have been suggested as design velocities by some of the companies surveyed. The velocities are expressed in units of feet per second (fps).

93

Wind and Swell	Approach Conditions	SHIP DISPLACEMENT		
		Up to 3,000 tons	Up to 10,000 tons	Over 10,000 tons
Strong	Difficult	2.5	2.0	1.5
Strong	Favorable	2.0	1.5	1.0
Moderate	Difficult	1.0	0.8	0.6

Actual conditions should determine what value is used for design. Future studies of the impact velocity might hinge on the human factors involved as well as quantifying the velocity relationships.

The energy equation has often been presented in the form:

$$E_f = 1/2 \; E$$

This indicates that the C value equals 1/2. If the individual factors are studied under common conditions, a slightly higher general value of C in indicated.

Case 1.
$$C = C_E \; C_S \; C_C \; C_H \tag{9}$$

C_E = 0.38 (Impact at sixth point)
C_S = 0.90 (General stiffness)
C_C = 0.80 (Closed pier structure)
C_H = 2.00 (B = 2D)

$$C = 0.46$$

Case 2.
$$C = C_E \; C_S \; C_C \; C_H \tag{9}$$

C_E = 0.50 (Impact at quarter point)
C_S = 0.90 (General stiffness)
C_C = 1.0 (Open structure)
C_H = 2.0 (B = 2D)

$$C = 0.90$$

For general use a value of C = 0.68 is recommended. This value is the average of the two cases illustrated above and appears more reasonable for design than the arbitrary assumption of C = 0.50. For most cases, however, evaluation of each coefficient factor is recommended.

The firm of Ewin, Campbell, and Gottlieb of Mobile, Alabama has provided us with four details of a fender study they did for the Alabama State Dock Department for berth numbers 3, 4, 5, 6, 7, and 8. These details appear on pp. 96, 97.

ALTERNATE DYNAMIC THEORY

General

As an alternate to the energy procedure just described, it is possible to assume that the pile system is equivalent to a mass supported by a spring. In such a system, it is assumed that the spring constant k is equivalent to the response of the pile and the spring mass M represents the ship. Examination of such a sprung mass system will result in the following general equations:

$$y_{max} = v_o/\lambda \qquad (1)$$

$$a_{max} = v_o \lambda \qquad (2)$$

$$P_{max} = k\, y_{max} \qquad (3)$$

$$t_{max} = \pi/2\,\lambda \qquad (4)$$

where:

$$\lambda = \sqrt{k/M} \qquad (5)$$

$$y = \text{displacement}$$

$$a = \text{acceleration}$$

$$P = \text{force}$$

$$t = \text{time}$$

Parameters

W = weight of ship (tons)

v_o = initial velocity of ship (knots)

Conversion: W_s (kips) = W x 2.

$$M = W_s/g = W_s/(32.2 \times 12) \qquad ksec^2/in$$

$$v_o = V_K x (1.689 \times 12) \qquad in/sec$$

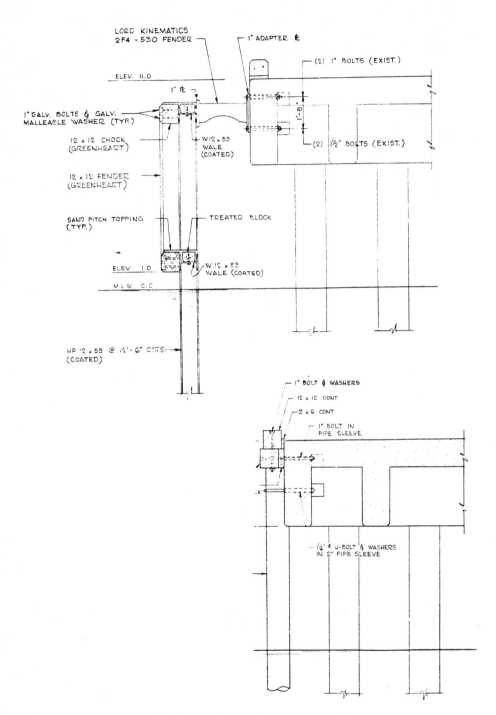

Various Schemes for Berth Conditions

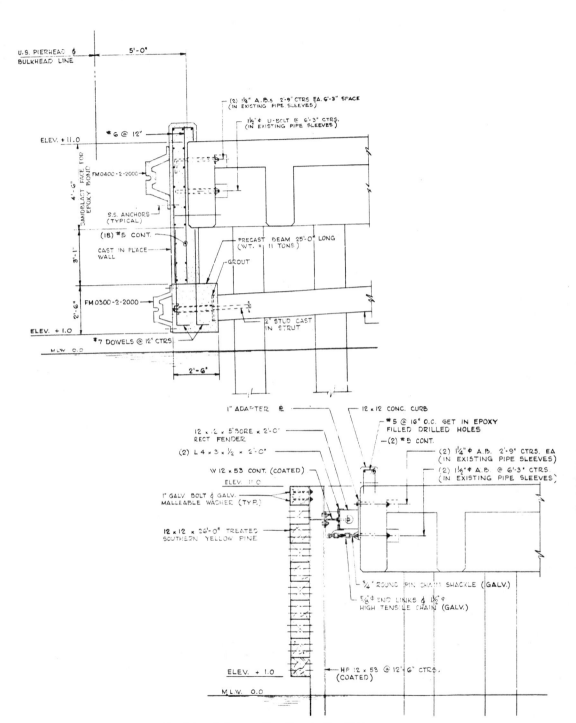

Various Schemes for Berth Conditions

The spring parameter k, which is equivalent to the pile deformation due to a unit load is computed as

$$k = 1/\Delta_p \qquad\qquad (5)$$

where:

$$\Delta_p = (\frac{1 \; x \; L^3}{3EI_p}) \; C$$

I_p = inertia of pile

E = modulus of elasticity

L = pile length

C = D.F. (explained in Chapt. VI)

Example

A large single birthing dolphin which is composed of 34 piles and a large rigid pile cap must absorb a ship of W = 250,000 tons and velocity v_o = 0.30 knots. Attached to the dolphin is a fender system which has a k curve defined as:

$$k_f = .028 \; \Delta^2 - 2.7\Delta + 87.9 \qquad\qquad (k/in)$$

Examination of the unit displacement of the dolphin, accounting for the soil interaction, gives the following:

$$k_d = 2300. \qquad (k/in)$$

Because these two systems will interact in series, the equivalent k is:

$$k_e = \frac{k_d \; . \; k_f}{k_d + k_f} \qquad , \; therefore$$

$$k_e = \frac{90 \; x \; 2300}{(90+2300)} = 86.6 \; k/in.$$

The ship parameters are converted to the proper units giving:

$$M = 250,000 \; x \; 2/(32.2 \; x \; 12) = 1450.5 \; \frac{ksec^2}{in}$$

$$v_o = 0.30 \; x \; 1.689 \; x \; 12 = 6.08 \; in/sec$$

Using equation (5);

$$\lambda = \sqrt{k/M} = (\frac{86.5}{1450.5})^{1/2} = 0.24$$

the maximum effects are therefore:

Deflection

$$y_{max} = v_o/\lambda$$

$$y_{max} = 6.08/0.24 = 25.33 \text{ in.}$$

Force

$$P_{max} = k \; y_{max}$$

$$P_{max} = 86.6 \times 25.33 = 2191.0^k$$

Acceleration

$$a = v_o \; \lambda$$

$$a = 6.08 \times 0.24$$

$$a = 1.45 \text{ in/sec}^2$$

Stopping Time

$$t = \pi/2\lambda$$

$$t = \pi/(2 \times 0.24)$$

$$t = 6.54 \text{ sec.}$$

The stress in the pile group can now be evaluated by first dividing the maximum force $P = 2191.0^k$ by the number of piles, which gives:

$$P_{pile} = \frac{2191.}{34} = 64.4^k$$

the moment at the end of the 70 ft. long pile is:

$$M = PL = 64.4 \times 70 = 4508^{k'}$$

The induced stress in the pile, which has a section modulus of $S = 1149.0 \text{ in}^3$ is therefore:

$$f = M/s = \frac{4508 \times 12}{1149}$$

$$f = 47 \text{ ksi}$$

CHAPTER V

HAND COMPUTATIONS

INTRODUCTION

This chapter contains several examples of simplified fender designs based on the design procedures and assumptions discussed previously. Each problem is presented in a step-by-step fashion with a range of solutions presented where applicable. It is noted that many facilities to which fenders are applied consist of pile groups. Where these piles act as a group, the design steps presented do not account for the distribution of the energy absorbed among the individual piles. Finally, the "factor of safety" concept may be applied to the solution of fendering requirements though the most desirable design method would involve solving the system for various conditions and selecting an appropriate design based on the results. Seven examples will be shown, namely;

1. A Timber Framework System

2. A Draped Rubber Fender

3. A Raykin Fender

4. A Pneumatic and Foam Filled Fender

5. A Lord Flexible

6. A Gravity Fender

7. A Steel Spring

EXAMPLE NO. 1: A TIMBER FRAMEWORK SYSTEM

Design a timber framework system for the given conditions:

1. A concrete pier with partially closed supporting structure.

2. With the shipping traffic consisting primarily of bulk ore carriers with average values of:

$$\text{Length} = 740'$$

$$\text{Breadth} = 101'$$

$$\text{Depth} = 56'$$

$$\text{Draft (Loaded)} = 38.5'$$

$$\text{Dead Weight} = 50,000 \text{ long tons}$$

$$\text{Displacement} = 62,500 \text{ long tons}$$

3. With an angle of approach $\approx 10°$ and the velocity ≈ 0.25 feet per second.

Solution:

I. Evaluate the coefficients.

a) Assume impact at quarter point $C_E = 0.5$

b) Partially open structure $C_C = 0.9$

c) Assume stiffness ratio $C_S = 0.9$

d) Assume the hydrodynamic coefficient which can be expressed by the following equation:

$$C_H = 1 + 2D/B = 1 + 2(38.5')/101' = 1.76$$

e) Solve for $C = C_E \cdot C_C \cdot C_S \cdot C_H = 0.71$

II. Determine the collision or berthing velocity.

$$V = 0.25 \text{ fps } (\cos 10°) = 0.2462 \text{ fps}$$

III. Determine the energy which must be absorbed by the fendering system.

$$E = C \frac{W}{2g} \cdot V^2 = 0.71 \left(\frac{62,500 \times 2240}{2 \times 32.2}\right)(0.246)^2$$

$$E = 93,900 \text{ ft. lbs.}$$

IV. Wale Design:

a) Assume the ship contacts two (2) fender piles.

b) Assume a wale size of 10" x 12" supported on 4" blocks, 10" on center.

c) Assume the amount of compression to equal 1/20 of the thickness.

$$d_1 = (1/20)(12" + 12") = 1.2"$$

d) Compute the amount of bending at the quarter points.

$$d_2 = \frac{Pa}{6EI}[\frac{31^2}{4} - 4a^2]$$

where:

1 = Length on center = 12" x 10' = 120 in.

a = 1/4 = 120/4 = 30"

E^* = Young's Modulus for given material

I = Moment of inertia $(bh^\#/12)$ = 1440 in.^{4}

* = The material properties for a given species of wood may be found by contacting the National Forest Products Association or consulting the Forest Products Manual.

If it is assumed that the waler is to be made of Douglas fir, the following properties would apply:

$$E = 1.90 \times 10^6 \text{ psi}$$

Impact bending = 4800 psi

Impact compression = 1000 psi

$$d_2 = \frac{P(30")}{6(1.9 \times 10^6)(1440 \text{ in.}^4)} [\frac{3(120")^2}{4} - 4(30")^2]$$

$$d_2 = 1.32 \times 10^{-5} P$$

e) Then equate the energy used in bending and compression to that absorbed by the fendering system.

$$2P(d_1 + d_2) = C \frac{W}{2g} V^2$$

$$2P(1.32 \times 10^{-5}P + 1.2) = 93,900 \text{ ft. lbs.}$$

$$P = 29,530 \text{ lbs.}$$

f) Compute the compression and bending stresses to insure that they are within the minimum values.

$$\text{Compression stress} = \frac{P}{\text{Bearing Area}}$$

Assume a 5" width of contact between fender pile and water:

$$\text{Compression} = \frac{29,530}{5" \times 12"} = 492 \text{ psi} < 1000 \text{ psi} \quad \text{O.K.}$$

$$\text{Bending} = M_c/I = P_{ac}/I = \frac{29,530(30)(5)}{1440} = 3076 \text{ psi} < 4800 \text{ psi}$$

$$\text{O.K.}$$

EXAMPLE NO. 2: A DRAPED RUBBER FENDER

Design a Draped Rubber Fender utilizing the same parameters as in Example No. 1.

Solution:

    I. Evaluate the coefficients.

$$C = 0.71$$

    II. Determine collision velocity.

$$V = 0.2462 \text{ fps.}$$

    III. Determine energy absorbed by fendering system.

$$E = 93,900 \text{ ft. lbs.}$$

Assume contact surface = 2 feet.

$$E/\text{contact} = 46,950 \text{ ft. lbs./ft.}$$

IV. Find appropriate fendering system from energy tables located in the Appendix using;

$$E = 93.9 \text{ K-ft. or } 1126.8 \text{ K-in.}$$

and

$$E/ft. = 46.95 \frac{\text{K-ft.}}{\text{ft.}} = 563.4 \frac{\text{K-in.}}{\text{ft.}}$$

The following are some fendering possibilities.

		Energy at Percent Deflection					
Company	Fender Name	50	60	70	80	90	100
	Goliath (K-in.)						
	1000 x 500 (K-in.)				1130	1433	2000
	1200 x 600 (K-in.)			1303	1738	2170	2610
	1400 x 800 (K-in.)				950	1390	2000	3475
	1500 x 800 (K-in.)				1000	1564	2260	3475
Goodyear	None Available						
Uniroyal	60" Cylindrical $(\frac{\text{K-in.}}{\text{ft.}})$					630	828

Any of the above fendering systems may be used. The choice among them may be based on economy, hull pressure, or installation requirements.

The computation of the hull pressure is as follows:

a) Assume a fender size: 1000 x 500

b) Determine the amount of deflection for given energy.

Since we need D = 1126.8 K-in. and at 80 percent deflection we have 1130 K-in., we are within the range. Therefore, from the energy table Δ_{max} = 20.5" and the deflection = 20.5" x 0.80 = 16.4".

c) Determine the stiffness coefficient from the appropriate curve.

At a 16.4" deflection, K = 8.5 Kips per inch

d) Determine the reaction force.

$$F = K \times \Delta = (8.5)(16.4) = 139.4 \text{ Kips}$$

e) Determine surface area of contact.

Assume approximately 1 sq. ft. or 144 sq. in.

f) Determine hull pressure.

$$\text{Pressure} = 139.4 \text{ K}/144 \text{ in.}^2 = 0.97 \text{ Kips/in.}^2$$

g) Check this against ships' specifications and it it checks,
then we use the assumed fendering system.

EXAMPLE NO. 3:   THE RAYKIN FENDER

Design a Raykin Rubber Fender given the same design parameters as in
the previous two examples.

Solution:

Since this is a rubber fender design, the same procedure and calculations
apply as in Example No. 2, and we go directly to the Raykin tables to choose
our system.

The following Raykin fenders are available for the mathematics of Example
No. 2.

Company	Size	Energy at Percent Deflection				Δ_{max}
		70	80	90	100	
General Tire and Rubber, Raykin	E-40 tons			950	1180	18.8 in.
	-45 tons			1080	1330	18.8
	-50 tons		936	1200	1520	18.8
	-60 tons		1074	1380	1740	18.8
	F-40 tons			1100	1332	22.4
	-45 tons		1010	1248	1500	22.4
	-50 tons		1120	1380	1632	22.4
	-60 tons	1020	1320	1660	2040	22.4

Hull pressure is tabulated for a bearing area of 2 sq. ft. or 288 sq. in. The procedure and results are tabulated below.

Size	Percent Deflection for E = 1126.8 K-in.	Deflection (Δ, in.)	K (K/in.)	Reaction Force (K x Δ, K)	Pressure (psi)
E-40 tons	97.8	18.4	6.2	114.1	396
-45 tons	92.0	17.3	7.0	121.1	420
-50 tons	87.3	16.4	8.8	144.3	500
-60 tons	81.8	15.4	9.4	144.8	503
F-40 tons	91.3	20.5	4.7	96.4	335
-45 tons	85.0	19.0	5.4	102.6	356
-50 tons	80.4	18.0	6.0	108.0	375
-60 tons	73.7	16.5	7.5	123.0	430

These pressures are all within the allowable limits for most cargo ships. The choice among them may be based on economics.

EXAMPLE NO. 4: PNEUMATIC AND FOAM FILLED FENDERS

Design a pneumatic or foam filled fender given the following conditions:

1. A concrete pier with a completely closed structure.

2. With the shipping traffic consisting primarily of tankers with average values of:

Length   = 1130'

Breadth   = 110'

Draft   = 65'

Dead Weight   = 330,000 long tons

Displacement = 380,000 long tons

3. With an angle of approach   20° and the velocity   0.25 feet per second.

Solution:

I. Evaluate coefficients.

a) Assume impact at quarter point, therefore, C_E = 0.5

b) Closed structure, therefore, C_C = 0.8

c) Assume stiffness ratio, therefore, C_S = 0.9

d) Assume the hydrodynamic coefficient which can be expressed by the following equation:

$$C_H = 1 + 2D/B = 1 + 2(65')/110' = 2.17$$

e) Solve for $C = C_E \cdot C_C \cdot C_S \cdot C_H = 0.78$

II. Determine the collision or berthing velocity.

$$V = 0.25 \text{ fps } (\cos 20°) = 0.235 \text{ fps}$$

III. Determine the energy which must be absorbed by fendering system.

$$E = C \cdot \frac{W}{2g} \cdot V^2 = \frac{0.78(380,000 \times 2240)(0.235)^2}{2 \cdot 322}$$

107

$$E = 569{,}350 \text{ ft. lbs.}$$

$$E = 6830 \text{ K-in.}$$

The following are some fender possibilities:

Company	Fender Name	Energy of Percent Deflection					Δ_{max}
		30	40	50	60	70	
Yokohama	Pneumatic						
	2500 x 5500			3,900	6,900		
	3300 x 6500		4,560	8,940			
	4500 x 9000	5,400	12,000	23,750	42,300		177.2
Seward	Foam Filled						
	8 x 16				6,000	9,300	96
	10 x 16		3,600	6,900	9,240		120
	10 x 20		4,500	7,500	11,400	18,000	120
	11 x 22		6,000	9,900	15,300	24,000	132
Sampson	Sprayed						
	8 x 16			4,200	7,320		96
	8 x 20			5,100	9,480		96
	10 x 16			6,600	11,500		120
	10 x 20		4,320	7,200	13,200		120

Any of the above fendering systems filfills the energy requirements. The choice among them may be based on economy, hull pressure, or installation requirements.

The computation of the hull pressure will compare one of each of the above types of fenders. The table below shows the significant values for energy absorption of 6830 K-in. K values are taken from the appropriate curves.

Company	Fender Name	Percent Deflection for 6830 K-in.	Deflection (in.)	K (K/in.)	Reaction Force (K)
Yokohama	Pneumatic				
	3300 x 6500	45.2	58.7	7.2	422.6
Seward	Foam Filled				
	10 x 16	50.0	60.0	3.8	228.0
Sampson	Sprayed				
	10 x 20	48.7	58.4	6.0	350.4

The Seward fender yields the lowest reaction force for this example and would, therefore, be the first choice if hull pressure is a significant design requirement. The bearing area to be used in the calculation is dependent upon the contours of the ship traffic.

EXAMPLE NO. 5: LORD FLEXIBLE FENDER

Design a Lord flexible fender for the given situation:

1. A concrete pier with an open structure.

2. With the shipping traffic consisting of passenger ships with average values of:

Length	= 920'
Breadth	= 105'
Draft	= 32'
Depth	= 45'
Dead Weight	= 10,500 long tons
Displacement	= 27,000 long tons

3. With an angle of approach 15° and the velocity 0.25 feet per second.

Solution:

I. Evaluate coefficients

   a) Assume impact at sixth point C_E ⬤ 0.38

b) Open structure $C_C = 1.0$

c) Assume stiffness ratio $C_S = 0.9$

d) Assume the hydrodynamic coefficient which can be expressed by the following equation:

$$C_H = 1 + 2D/B = 1 + 2(32')/105' = 1.61$$

e) Solve for $C = C_E \cdot C_C \cdot C_S \cdot C_H = 0.55$

II. Determine the collision or berthing velocity.

$$V = 0.20 \text{ fps } (\cos 15°) = 0.193 \text{ fps}$$

III. Determine the energy which must be absorbed by fendering system:

$$E = C \cdot \frac{W}{2g} \cdot V^2 = \frac{0.55(27,000 \times 2240)(0.193)^2}{2 \cdot 32.2}$$

$$E = 19,240 \text{ ft. lbs.}$$

$$E = 230.9 \text{ K-in.}$$

The following are some fender possibilities obtained from the Lord tables

Company	Fender Name	Energy at Percent Deflection					Δ_{max}
		60	70	80	90	100	
Lord	2F4-212				228	264	14.5
	-319		215	275	325	390	14.5
	-390	230	275	350	410	460	14.5

The hull pressure is based on a bearing area of one square foot or 144 square inches. The K values are obtained from the appropriate curves.

Company	Fender Name	Percent Deflection	Deflection (in.)	K (K/in.)	Reaction Force	Pressure (psi)
Lord	2F4-212	90.6	13.1	1.75	22.9	159
	-319	72.5	10.5	3.05	32.0	222
	-390	60.0	8.7	4.10	35.7	250

The above pressures are within allowable limits for most passenger ships. The choice among them may be based on economics or installation limitations.

EXAMPLE NO. 6: GRAVITY FENDER

Design a gravity fender utilizing the same parameters as Example No. 5. Solution:

    I.   Evaluate the coefficients.

$$C = 0.55$$

    II.   Determine the collision or berthing velocity.

$$V = 0.193 \text{ fps}$$

    III.   Determine energy absorbed by the fendering system.

$$E = 19,240 \text{ ft. lbs.}$$

$$E = 230.9 \text{ K-in.}$$

    IV.   Structural limitations restrict the amount of deflection to 24 inches.

The equation of the energy loss for the gravity fender is;

$$E = F \times d$$

where E is energy loss, F is the weight of the fender, and d is the horizontal displacement. Rearranging the terms yields;

$$F = \frac{E}{d} = \frac{230.9 \text{ K-in.}}{24 \text{ in.}} = 9.6 \text{ Kips}$$

The structure must then be built to support this weight and allow the required displacement of the mass.

EXAMPLE NO. 7: STEEL SPRING FENDER

Design a steel spring fender utilizing the same parameters as Example No. 5.

Solution:

I. Evaluate the coefficients.

$$C = 0.55$$

II. Determine the collision or berthing velocity.

$$V = 0.193 \text{ fps}$$

III. Determine the energy absorbed by the fendering system.

$$E = 19,240 \text{ ft. lbs.}$$

$$E = 230.9 \text{ K-in.}$$

The equation of the energy loss for the steel spring is;

$$E = K \times d^2$$

where E is the energy loss, K is the spring constant, and d is the displacement along the length of the spring. Rearranging terms yields;

$$K = E/d^2$$

If structural limitations restrict displacement to ten inches, then;

$$K = \frac{E}{d^2} = \frac{230.9 \text{ K-in.}}{(10 \text{ in.})^2} = 2.309 \text{ K/in.}$$

The design of the spring to meet these requirements may follow any structural method.

CHAPTER VI

DESIGN APPLICATION AND COMPUTERS

INTRODUCTION

The design of fendering systems which protect bridge piers against ship impact or at minimum retard direct collision has utilized basic fundamentals of physics and simple pile equilibrium as illustrated in the previous two chapters. Such assumptions can often lead to grossly overdesigned systems or in some cases inaccurate force evaluation in the entire support system. Therefore, this chapter deals with the development of improved design criteria and the application of a complex computer oriented dynamic solution.

THEORY

General Techniques. - The general response of a piling system, when subjected to a ship, is computed by removing the pile and examining its effect as a cantilever beam, as shown in Figure 1. The interaction of lateral elements, such as walers are neglected and thus a conservative design. Two general theoretical equations are used by the designer, and are based on force-acceleration and kinetic energy relationships.

Force-Acceleration:

the induced or applied force to the system, caused by the ships impact is;

$$F_a = M(v_i^2 - v_f^2)/2\Delta_s \tag{1}$$

where: M = mass of ship

Δ_s = deformation of system at point of impact

v_i, v_f = initial and final velocity

the resisting force of the system is:

$$F_r = 3\Delta_s E(I/D.F.)/L^3 + \Sigma k\Delta_s \tag{2}$$

where: E = modulus of elasticity of pile

I = inertia of pile

D.F. = lateral distribution of load due to lateral stiffness or effect

L = cantilever length of pile

k = spring constant of fendering

the induced moment and stress is computed from;

$$M = F_a \times L \text{ and} \tag{3}$$

$$f = M/(S/D.F.) \tag{4}$$

where: S = section modulus.

115

In applying this method the designer would assume on allowable Δ_s and initial stiffness I. If the resisting force $F_r > F_a$, then the actual Δ_s would be smaller than assumed. The induced stress f would be compared to the allowable or ultimate stress of the material.

Kinetic Energy:

The induced energy caused by the ship is given by:

$$E_{in} = \frac{1}{2} M v_i^2 (C_H)(C_s)(C_c)(C_E) \qquad (5)$$

where: v_i = initial or translational ship velocity

C_H = hydrodynamic coefficient = $1 + \frac{2D}{B}$

D = draft of ship

B = beam of ship

C_E = eccentricity coefficient

C_s = softness coefficient

C_c = configuration coefficient

the C coefficients (C_E, C_s and C_c) can be set equal to 1.0, for the worst case. Other variations can be obtained for specific ship variables, as given in Ref. (2).

The output energy or that energy that can be absorbed by the piling system is;

$$E_o = F \cdot \Delta_p + \Sigma \frac{1}{2} k \Delta_f^2$$

but $\Delta_p = FL^3/3E(I/D.F.)$, therefore

$$E_o = F^2 L^3/(3EI/D.F.) + \Sigma \frac{1}{2} k\Delta^2 \qquad (6)$$

Using equations (5) and (6) and assuming $\Delta = FL^3/3EI/D.F.$ or zero, the induced force F is determined. The resulting Δ can then be evaluated and used to reevaluate E_o if $\Delta = 0$ was originally assumed. The resulting moment and stress is found as per equations (3) and (4).

System Technique. –

A complete pile system is shown in Figure 2, and includes the support piles and lateral walers, excluding fenders. This system in effect is a cantilever grid plate, subjected to a lateral load. The response of such a system has been determined by using a matrix formulation as represented by a finite difference scheme (3). The stiffness of this grid is defined as D_x and D_y, which is computed as $D_x = EI_x/\lambda y$ and $D_y = EI_y/\lambda x$.

Such a scheme, which has been computerized, has permitted direct solution of the deformations and forces throughout the pile grid system. Relationship between the induced system deformation and isolated pile gives a new design technque, classified herein as load distribution factor. This factor shows what percentage of load goes to each pile when the load is applied to one pile.

PARAMETRIC STUDY

In order to determine the distribution factor, is necessary to determine the system response. The determination of this factor (DF) has been obtained for typical grid stiffnesses (D_x, D_y) and span length (L) or height of the pile. A unit load effect was used in examining the system and single pile.

Longitudinal Stiffness. – D_y

The range in the stiffness $D_y = \dfrac{EI_y}{\lambda x}$, was determined by examination of typical steel HP, steel W, and 12 in. – 28 in. round timber members which are used in piling systems. The spacing λx was varied between 5 ft. and 25 ft., in 5 feet increments. The length of cantilever plate was varied between 20 ft. and 60 ft. in 4 feet increments.

117

<u>Transverse Stiffness.</u> - D_x

The range in the stiffness $D_x = \frac{EI_x}{\lambda y}$, was determined by also ex-

amining typical walers, which consisted of steel W and 10 in., 12 in.,

and 14 in. timber sections. The spacing λy was varied the same as

the λx variable.

<u>Range in Parameters.</u>

A study of all of the resulting stiffnesses indicates the follow-

ing ranges;

Variable	Lower Bond	Upper Bond
D_y	2×10^4 k-in.	6×10^5 k-in.
D_x	4×10^3 k-in.	10×10^4 k-in.
L	20 ft.	60 ft.

<u>Grid Difference Solutions</u>

Using these ranges in parameters and applying a unit load, the

maximum deformation in the system has been obtained. In all these

solutions a maximum of ten vertical mesh lines was used, where the

spacing of the lines was set equal to a constant $\lambda x = 60$ in., which

gave a width of 45 ft. The mesh points along these vertical lines was

fixed at $\lambda y = 48$ in. for the range of 20 ft. to 60 ft.

The solution of systems has given the Δsys which was then di-

vided by the factor $L^3/3EI_y$, which is called Distribution Factor (D.F.).

These results were then plotted, (D.F.) vs. pile height L, as given in

Figures 3 through 10. This ratio, given D.F., will now be described.

<u>Distribution Factor.</u>-

The finite difference cantilever grid equations can provide direct

deformation values along any pile. Depending on the lateral stiffness

(D_x), the deformation at the top of free edge of the piles can vary

dramatically. This variation is quite important if the designer wishes

118

to properly identify the interaction between the piles and an isolated pile. A convenient method to describe such interaction is to relate the deformation of the systems (Δs) to that of the isolated pile (Δp), which is called a distribution factor or:

$$D.F. = \frac{\Delta s}{\Delta p} \tag{7}$$

where: Δs = maximum deformation in the grid system (finite differences)

$$\Delta p = PL^3/3EI_y - \text{cantilever pile deformation}$$

Equation (7) signifies the reduction in deformation of an isolated pile when that pile is part of the system and thus the influence of lateral stiffness. Therefore, the stiffness (I) of an isolated pile can be increased by the amount of 1/D.F. or I sys = I_p/D.F. This factor has been referenced in equations (2) and (6).

The resulting distribution factors, for various relative stiffnesses, are given in Figures 3 through 10, and can be used for direct design.

COMPUTERIZED SOLUTION

As described previously, the system response can be evaluated by use of (D.F.) curves. This approach, although sufficient for engineering design, can be improved by utilization of structural dynamics and computerization. Such a system has been developed and incorporates the following features;

General.-

The interaction of a protective bridge fendering system, or wharfs which can consist of a series of piles imbedded in a soil medium and stiffened laterally, with possible absorbing devices attached, is a complex system under static conditions and highly redundant. This action is further complicated if the system is subjected to dynamic forces.

Another system which is utilized for bridge protection, is a cluster of piles wrapped with cables, or a cell dolphin or a series of steel piles with a rigid cap. The dynamic response of any of these systems is also complex.

In the past, the solution of either the fendering system or dolphin, has been examined by the engineer as a single element, fixed at the base (cantilever) and then applied as a basic physics relationship. This method rewards the engineer with simplicity but inherently may not be conservative nor safe. This condition has therefore led to the impetus of developing a computer oriented solution of such systems which can incorporate many variables, which are not possible in the simplified technique, and thus provide rapid and accurate solutions to a complex problem.

Assumptions.-

1. The piling interaction with the soil medium is considered, i.e., flexible supports.

2. Soil may be layered.

3. Piling group is considered as a three dimensional unit.

4. Interactions of the horizontal walers is considered.

5. Forces and deformations throughout all piles, at any time interval, can be evaluated.

6. The forces and deformations are evaluated along the length of each pile.

7. Rigid wharfs, fenders, dolphins, or combinations can be considered.

8. During ship impact, any pile that fails, is noted and the system is reevaluated.

9. Total energy in the system, input and output, is computed during each time interval.

10. The system may have any general plan orientation; i.e., (straight, curved etc.).

Input Data.-

The input requirements for the fendering or dolphin is flexible, depending on the problem. If the system is straight, then the spacing and point of impact need only to be specified. In the case of straight and curved systems, then the requirements are more rigorous. A sample of the various problems that can be investigated by the program are given in Sample Problems 1 through 10 for fendering system and Sample Problems 1 through 5 for dolphins.

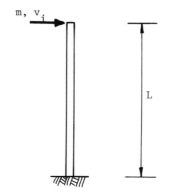

FIGURE 1

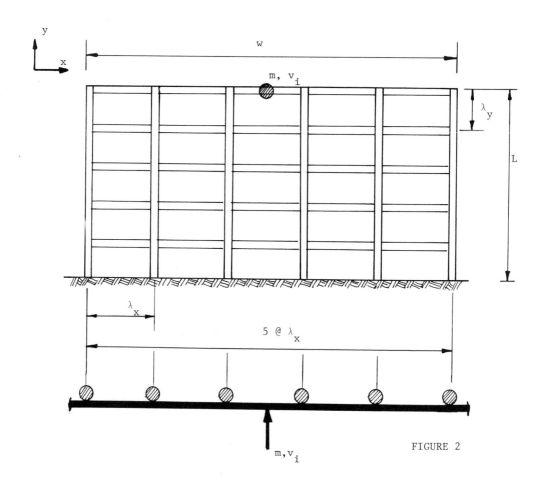

FIGURE 2

122

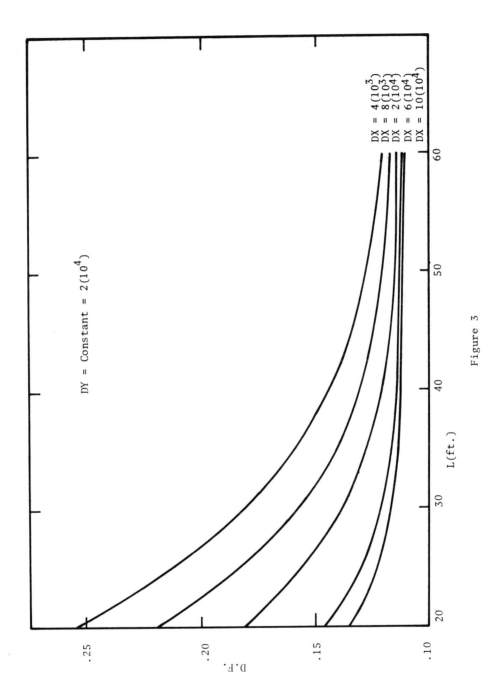

Figure 3

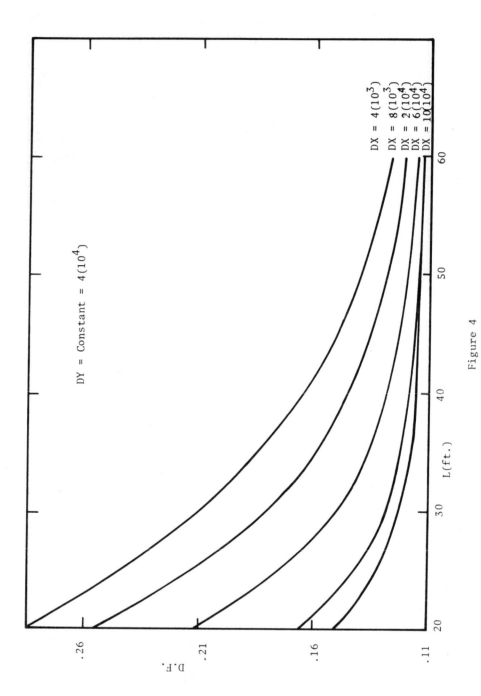

Figure 4

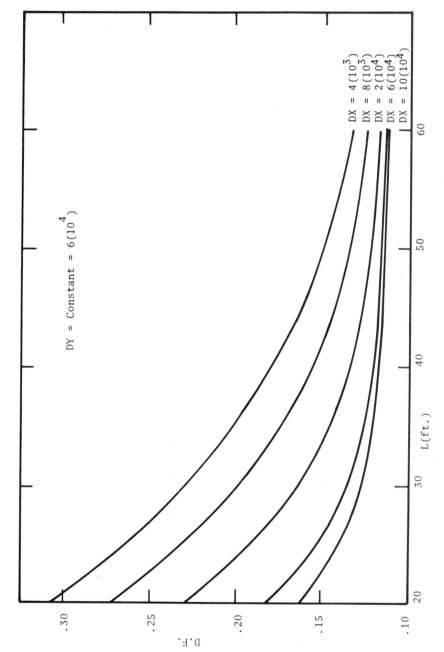

DY = Constant = $6(10^4)$

DX = $4(10^3)$
DX = $8(10^3)$
DX = $2(10^4)$
DX = $6(10^4)$
DX = $10(10^4)$

L(ft.)

D.F.

Figure 5

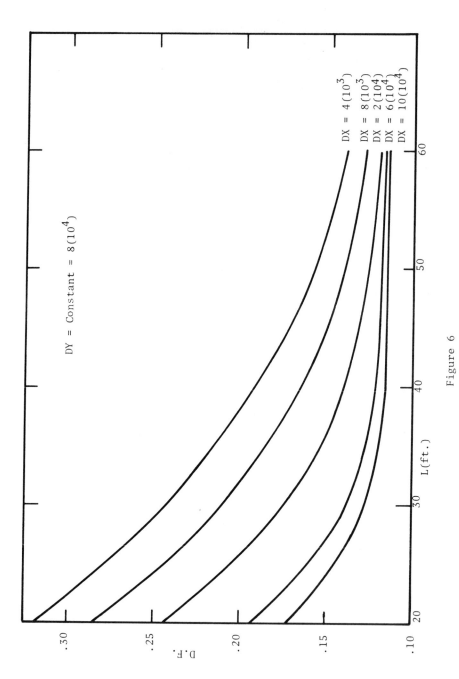

Figure 6

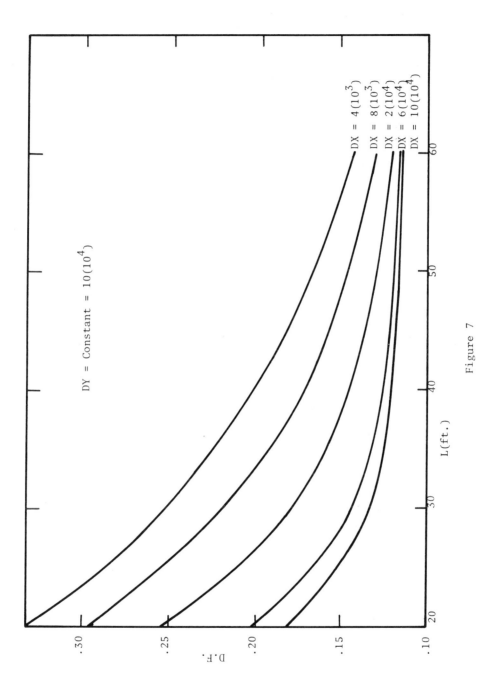

Figure 7

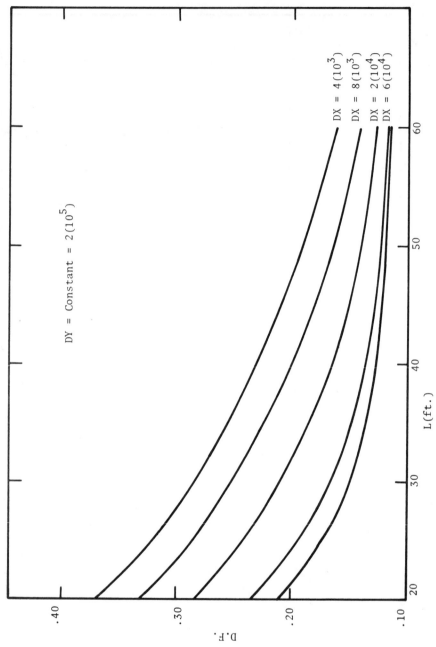

Figure 8

DY = Constant = $2(10^5)$

DX = $4(10^3)$
DX = $8(10^3)$
DX = $2(10^4)$
DX = $6(10^4)$

L(ft.)

D.F.

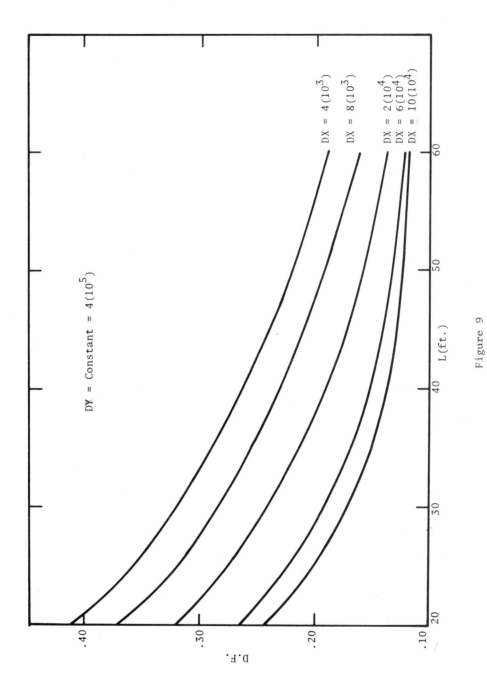

Figure 9

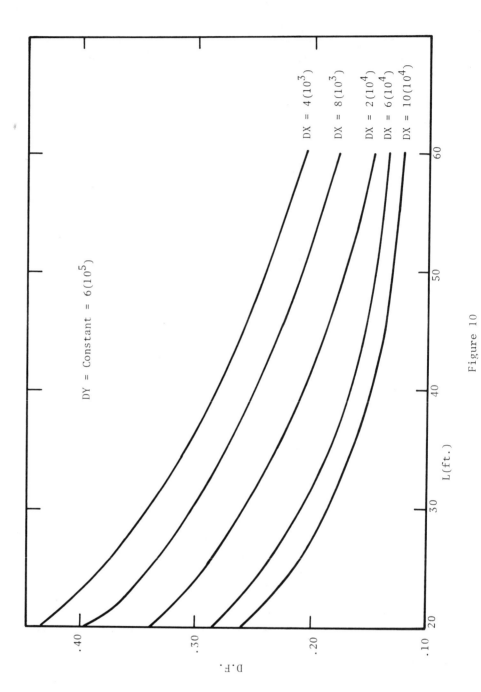

Figure 10

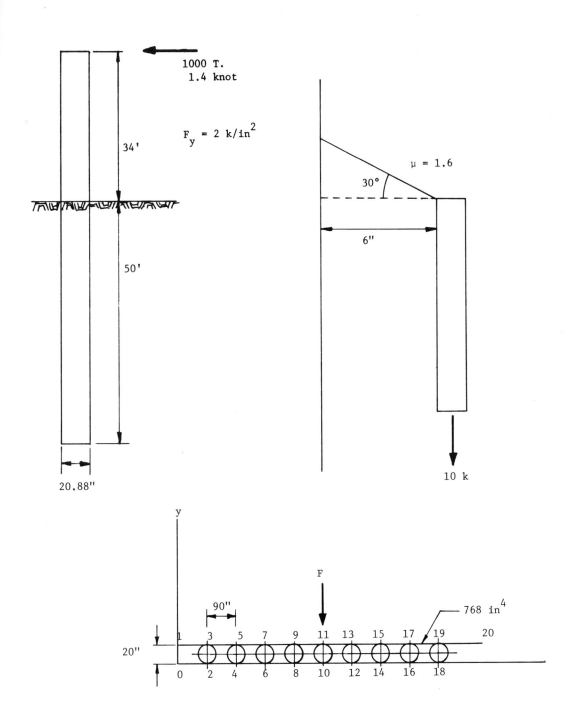

1000 T.
1.4 knot

$F_y = 2 \ k/in^2$

34'

50'

20.88"

$\mu = 1.6$

30°

6"

10 k

y

F

90"

768 in^4

20"

1 3   5   7   9   11   13   15   17   19   20

0   2   4   6   8   10   12   14   16   18

SAMPLE PROBLEM 1

131

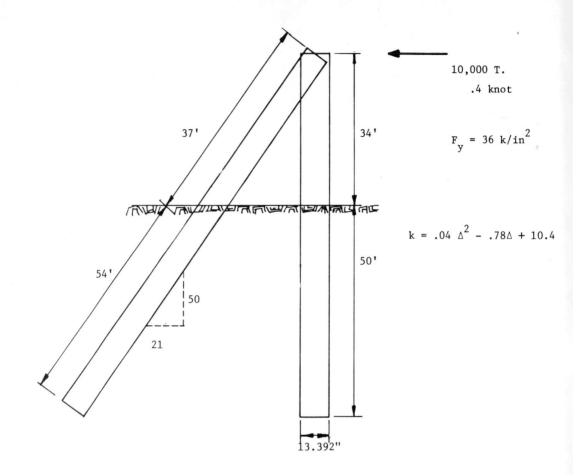

10,000 T.

.4 knot

$F_y = 36$ k/in^2

$k = .04 \Delta^2 - .78\Delta + 10.4$

37'

34'

54'

50

21

50'

13.392"

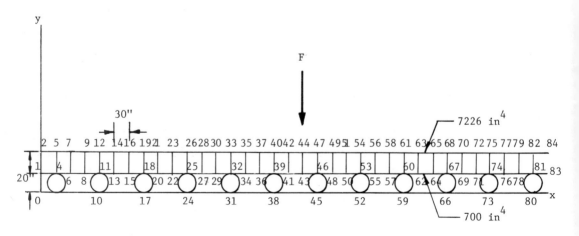

SAMPLE PROBLEMS 2, 8

132

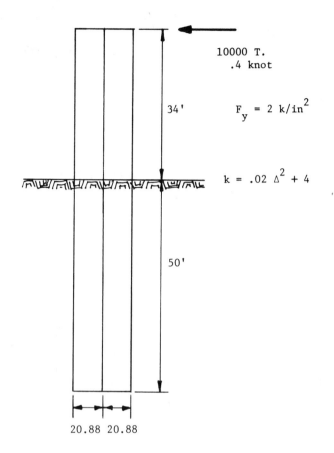

10000 T.
.4 knot

$F_y = 2 \text{ k/in}^2$

34'

$k = .02 \Delta^2 + 4$

50'

20.88 20.88

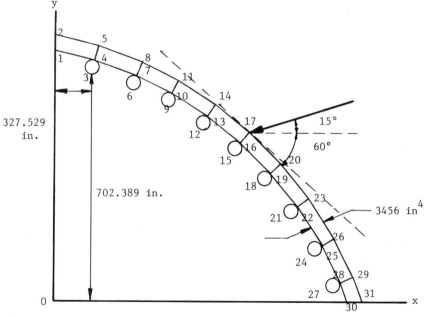

327.529
in.

702.389 in.

15°

60°

3456 in^4

133

SAMPLE PROBLEM 3

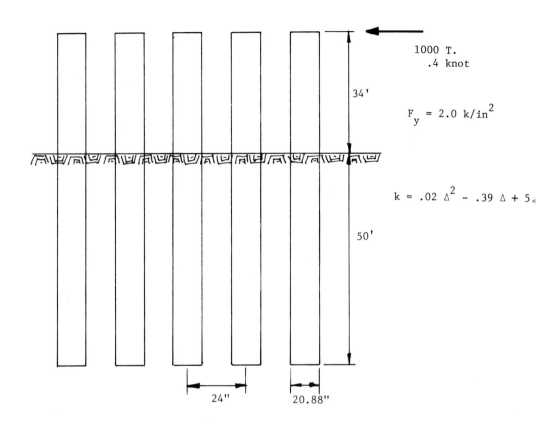

1000 T.
.4 knot

$F_y = 2.0 \text{ k/in}^2$

$k = .02 \, \Delta^2 - .39 \, \Delta + 5.$

34'

50'

24"   20.88"

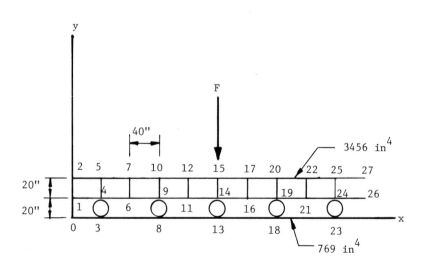

SAMPLE PROBLEMS 4, 9

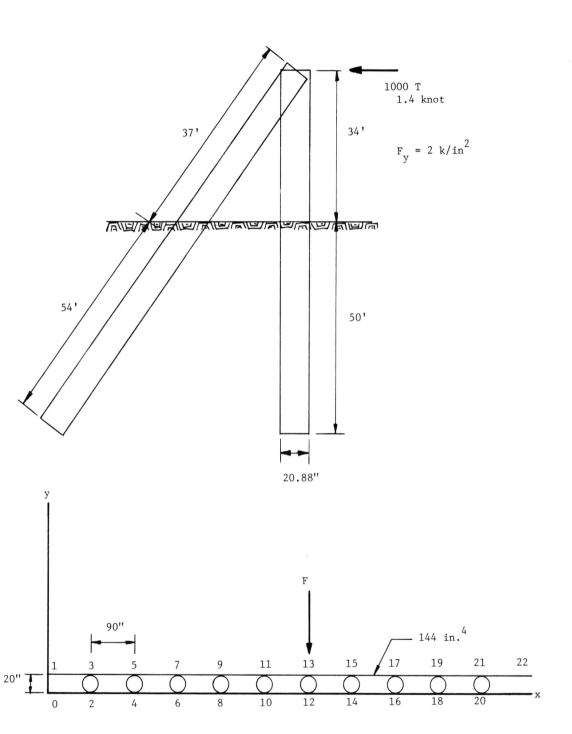

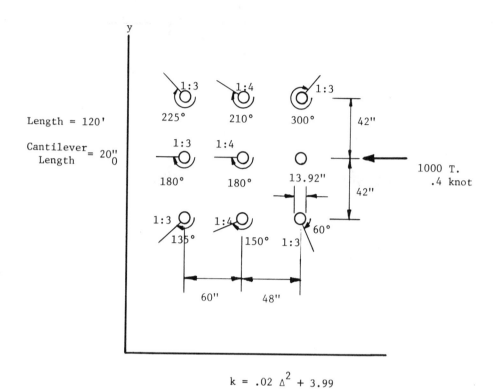

Length = 120'

Cantilever Length = 20"

225° 1:3   210° 1:4   300° 1:3   42"

1:3   1:4   1000 T.
180° 180°   13.92"   .4 knot

42"

1:3   1:4   60°
135° 150° 1:3

60"   48"

$$k = .02 \ \Delta^2 + 3.99$$

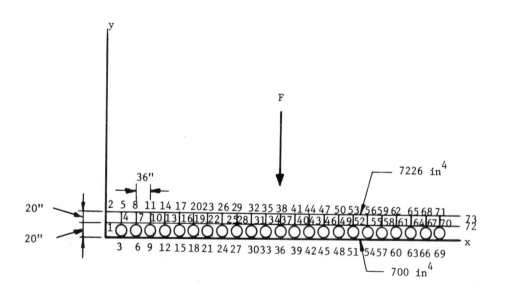

F

36"

20"

7226 in^4

2 5 8 11 14 17 20 23 26 29 32 35 38 41 44 47 50 53 56 59 62 65 68 71
4 7 10 13 16 19 22 25 28 31 34 37 40 43 46 49 52 55 58 61 64 67 70   73
1   72

20"

3 6 9 12 15 18 21 24 27 30 33 36 39 42 45 48 51 54 57 60 63 66 69

700 in^4

136

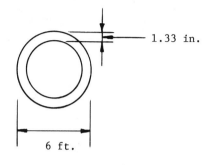

1.33 in.

6 ft.

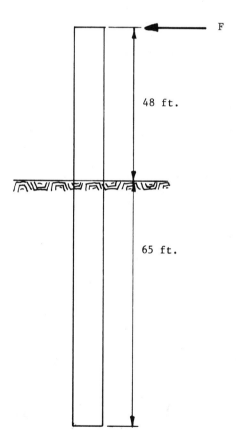

F

48 ft.

65 ft.

SAMPLE PROBLEM 1

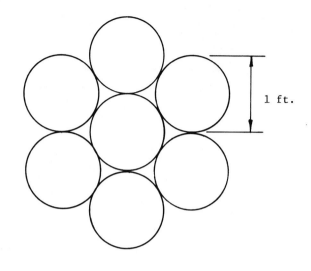

1 ft.

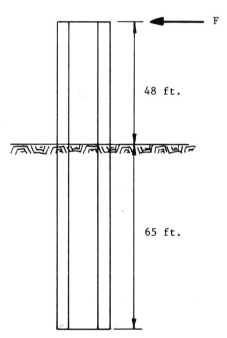

F

48 ft.

65 ft.

SAMPLE PROBLEM 2

138

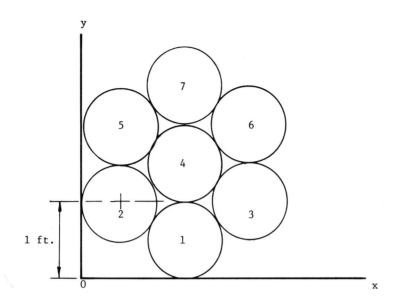

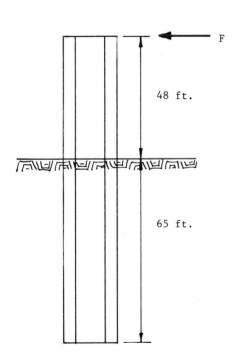

1 ft.

F

48 ft.

65 ft.

SAMPLE PROBLEM 3

139

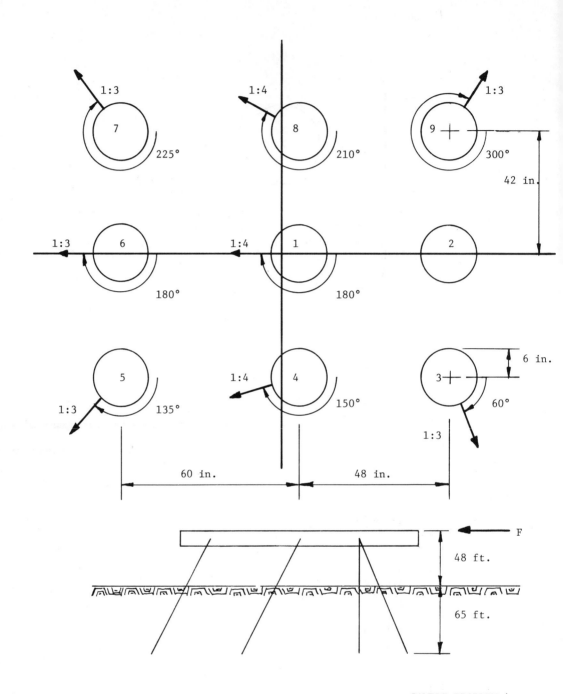

SAMPLE PROBLEM 4

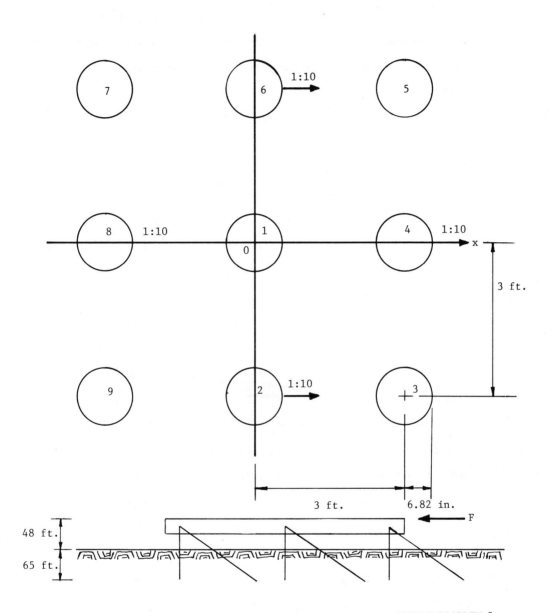

SAMPLE PROBLEM 5

141

Example 1

In order to demonstrate the use of the distribution factor (D.F.) as applied to pile supported fenders, the following wooden system will be examined, as shown in Figure 11. The parameters relative to this system are;

Length of pile (1) = 34 ft.

Pile Spacing (λx) = 80 in.

Width of system = 720 in.

Modulus of Elasticity E = 1600.ksi (pile & walers)

I pile = .0491d^4 = 19,180.in^4 (25"ϕ)

I waler = $\frac{1}{12}$ bd^3 = 833.in^4 (10"x10")

Fenders not used (K = 0.0)

The ship parameters are;

Weight of Ship = 10,000 ton (20,000. kip)

Velocity of ship (v_i) = 0.40 knot (8.21 in/sec)

The general pile system will be examined under three variations in lateral stiffness.

Case I will have only one waler at the top of the pile (λy = 34 ft.), the second condition Case II, the walers will be spaced at (λy = 5'-7"), and the last Case III, will have walers at λy = 1.0 ft. intervals.

The solutions for each of these cases are as follows;

CASE I

System of vertical piles with no fenders, shown in Fig. 11, with only one waler at the top. The plate is modeled by smearing the stiffness over the total area (D x L), which gives;

$$DY = \frac{(EI_y)pile}{\lambda_x} = \frac{(1600 \text{ k/in}^2) (19,180.\text{in}^4)}{(720/9)\text{in}} = 3.84 \times 10^5\text{k-in (use } 4 \times 10^5)$$

$$DX = \frac{(EI_x)_{waler}}{\lambda_y} = \frac{(1600 \text{ k/in}^2)(833 \text{ in}^4)}{34(12) \text{ in}} = 3.3 \times 10^3 \text{ k-in} \quad (\text{use } 3 \times 10^3)$$

from Figure 9, using $L = 34$ ft., $DY = 4.0 \times 10^5$ and $DX = 3 \times 10^3$, then

D.F. = .300. With this information the induced stresses, as computed by

the two methods are computed as follows;

    i) Force Acceleration:

        Apply Eq. (1), and assuming $\Delta s = 20$ in., gives;

Applied Force

$$F_a = M(v_i^2)/2\Delta s$$

$$F_a = (\frac{20,000.}{32.2}) \; \frac{(8.21)^2}{2(20/12)} \times \frac{1}{144} = 87.2^k$$

Resisting Force

$$F_r = 3\Delta s E(I/D.F.)/L^3 + \Sigma k\Delta s$$

$$F_r = 3(20) \; (1600.) \; (\frac{19,180}{.300})/(34 \times 12)^3$$

$$F_r = 100.2^k$$

$$F_r > F_a \; \therefore \; OK$$

Moment

$$M = F_a L = 87.2 \times (34 \times 12) = 35,577.6 \text{ k-in.}$$

Stress

$$\sigma = M/(S/D.F.), \text{ where } S = \frac{19,180.}{12.5} = 1534. \text{ in}^3, \text{ D.F.} = .300$$

$$\sigma = \frac{35577.6}{1534.} \times .300$$

$$\sigma = 6.9 \text{ ksi} < \sigma_u = 7 \text{ ksi (assumed)}$$

without D.F., $\sigma = 21$ ksi

ii) Kinetic Energy: Applying Eq. (5) the induced energy is computed as;

$$E_{in.} = \frac{1}{2} M v_n^2 \, (C_H)(C_S)(C_C)(C_E)$$

$$E_{in.} = \frac{1}{2} \left(\frac{20,000.}{32.2}\right) \frac{(8.21)^2}{144} (1.825)\,(1)\,(1)\,(1)$$

where: $C_H = 1 + \frac{2D}{B}$, D = 33 ft., B = 80 ft.

$C_H = 1.825 \quad C_S = 1,\ C_C = 1,\ C_E = 1$

therefore: $E_{in.} = 181.5$ k ft = 2178. k-in.

The output energy of the system, given by Eq. (6), is;

$$E_o = F^2 L^2/(3EI/D.F.)$$

$$E_o = F^2(12\times34)^2/(3\times1600.(19180./.300))$$

$$E_o = F^2\,(.221)$$

Equation $E_o = E_{in}$ gives

$$2178. = F^2 x\ .221$$

$$F^2 = 9855.2$$

$$F = 99.3\ k$$

Moment:

$M = F \times L = 99.3 x(34x12) \quad 0,514.4 = k\ in$

Stress:

$$\sigma = M/(S/D.F.) = \frac{40,514.4}{(1534/.30)} = 7.9\ ksi > 7.0\ ksi$$

Marginal

CASE II

This case is similar to Case I, except the walers are now placed at 5'-7" intervals. This will provide a change in the Dx stiffness, which is:

$$D_x = \frac{(EI_x)waler}{\lambda y} = \frac{(1600)\,(833.)}{67} = 1.98\times10^4\ k\text{-}in$$
$$(\text{Use } 2\times10^4)$$

$$D_y = \frac{(EI_{y\ pile})}{\lambda y} = \frac{(1600)\ (19,180.)}{720} = 3.84 \times 10^5 \text{ k-in}$$

$$(\text{Use } 4 \times 10^5)$$

Examining Figure 9, gives D.F. = .220

   i) Force Accelerations

       As given in CAse I, $F_a = 87.2^k$,

       the resisting force is computed as;

$$F_r = 100.2(.300/.220) = 136.6^k$$

$$F_r > F_a \quad \therefore \text{ Section OK.}$$

       the induced stress is;

$$\sigma = 6.9(.220/.300) = 5.06 \text{ ksi} < 7.0 \text{ ksi} \quad \therefore \text{ OK.}$$

  ii) Kinetic Energy:

$$E_{in} = 2178. \text{kin.}$$

$$E_o = F^2(.221)\ (.220/.300) = .162F^2$$

       Equating $E_o = E_{in}$ gives the induced force F of;

$$2178. = F^2 \times .162; \quad F = 116.^k$$

       the induced stress is therefore;

$$\sigma = 7.9((.22/.30)\ (116./99.3) = 6.8 \text{ ksi} < 7.0 \text{ ksi}$$

CASE III

    This final case consists of walers spaced at 1.0 ft. intervals,

thus

$$D_x = \frac{(EI_x)\text{walers}}{\lambda y} = \frac{(1600)\ (833.)}{12} = 1.1 \times 10^5 \text{k-in.}$$

$$(\text{Use } 1 \times 10^5 \text{k-in})$$

$$D_y = 3.84 \times 10^5 \text{k-in (Use } 4 \times 10^5 \text{k-in)}$$

Using Figure 9, D.F. = .15

i) Force Acceleration

As given in Case I, $F_a = 87.2^k$

the resisting force is computed as;

$$F_r = 100.2(.300/.15) = 200.4^k$$

$$F_r > F_a \quad \therefore \text{ Section OK}$$

The induced stress is:

$$\sigma = 6.9(.150/.300) = 3.45 \text{ ksi} < 7.0 \text{ ksi}$$

ii) Kinetic Energy

$$E_{in} = 2.78.0 \text{ k-in.}$$

$$E_o = F^2(.221) \ (.150/.300) = .110 \ F^2$$

equating $E_o \neq E_{in}$ gives the induced force F;

$$2178 = F^2 \text{ x } .110$$

$$F = 140.7^k$$

The induced stress is therefore;

$$\sigma = 7.9(.15/.30) \ (140.7/99.3) = 5.6 \text{ ksi} < 7.0 \text{ ksi}$$

A comparison of the results are shown in Table 2, and show the effect the walers have on the reduction of stress on the piles, through lateral distribution of load. Also shown in the table are the results obtained from a computer simulations model (1).

146

TABLE 1

CASE	λx(in)	λy(in)	$D_y \times 10^5$(k-in)	$D_x \times 10^3$(k-in)
1.	90.	408.	4.0	3.0
2.	90.	67.	4.0	20.
3.	90.	12.	4.0	100.

TABLE 2

| | METHOD | | | | COMPUTER | |
| | Force Acceleration | | Kinetic Energy | | | |
CASE NO.	Force(k)	Stress(ksi)	Force(k)	Stress(ksi)	Force(k)	Stress(ksi)
1.	87.2	6.9	99.3	7.9	139.7	---
2.	87.2	5.06	116.	6.8	107.	---
3.	87.2	3.45	140.7	5.6	85.0	---

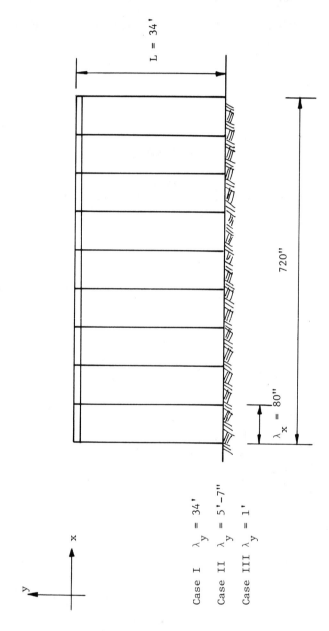

FIGURE 11

148

EXAMPLE 2

The docking of vessels in a port facility requires a structural system which can support the impact of the vessel, continue to be serviceable, have easy maintenance at a minimum cost.

In general, the warf or dock contains a front battered pile with possible walers and absorbing devices, as shown in Figure 12. In many instances, bulbus bow ships will shear off the bottom of the front pile, without any noticeable damage to the top of the pile until an inspection is made.

Thus, it is apparent that a new structural system or systems should be devised to remedy this problem and contain the following features:

1. Minimum Construction

2. Minimum Cost

3. Easily Maintained

4. Structurally Sound

5. Reasonable Life

This criteria can be met with improved technology and engineering judgment, an example of which will be described herein.

Problem

One of the major problems of a port, as mentioned previously, is the shearing off of pilings due to various types of ships. The design of a proper system to eliminate such a condition, and to provide a reasonable life to the system, will be given by considering a front vertical piling with a suspended wooden bulkhead with attached absorbing device. This example is only given to illustrate what may be done and how it may be technically accomplished.

The general plan of the system is shown in Figure 13, and consists of a series of vertical piles with a suspended wooden bulkhead and attached fenders. The bulkhead is supported laterally by a series of channels, Fig. 14, which slide along the W shape flange. Repair and removal of the bulkhead can easily be performed by maintenance personnel.

Solution

a) Analytical Solution

The response of the pile system will be determined by utilization
of a series of recently developed distribution factor (D.F.) curves
which are based on plate and girder theories and consider lateral interaction of the walers or grid. This interaction is incorporated in
the design by evaluation of the longitudinal (D_y) and transverse (D_x)
stiffness of the system, as shown in Figure 15.

As shown in Figure 14, the bulkhead is assumed to consist of
fifteen 6" x 12" treated wooden horizontal members spaced at 2'. The
vertical members comprise six 12" x 12" treated wooden members spaced
at 2'. The stiffness of this unit is computed as follows:

D_y:

$$D_y = EI/\lambda x = \frac{1600. \text{ x } 1728}{10 \text{ x } 12} = 23040. \text{ K-in}^2/\text{in.}$$

$$\text{Use } (2 \text{ x } 10^4 \text{ K-in})$$

$\lambda x = 10'$

$E = 1600. \text{ ksi}$

$I = \frac{1}{12} (12)(12)^3 = 1728 \text{ in}^4$

D_x:

$$D_x = EI/\lambda y = \frac{1600. \text{ x } 1728}{60 \text{ x } 12} = 3840. \frac{\text{K-in}^2}{\text{in.}}$$

$$\text{Use } (4 \text{ x } 10^3 \text{ K-in})$$

$\lambda y = 60'$

$E = 1600. \text{ ksi}$

$I = \frac{1}{12} (12)(12)^3 = 1728 \text{ in}^3$

151

using now the general energy equation

$$F = M_v{}^2/2\Delta \text{ gives:}$$

$$F_a = M_{v_i}{}^2/2\Delta_s$$

Assume $\Delta_s = 20"$, and for a ship W = 10,000 tons and $v_i = 0.40$ knots, then the applied force is;

$$F_a = (\frac{20000.}{32.2}) \ \frac{(8.21)^2}{2 \times 20/12} + \frac{1}{44} = 87.2^K$$

The resulting force of the steel pile is:

$$F_r = 3\Delta_s \ E \ (I/D.F.)/L^3 + K\Delta_s$$

assuming no fenders then K = 0 and $F_r = F_a$, gives;

$$87.2 = 3\Delta_s \ E \ (I/D.F.)/L^3, \text{ solving for I and letting}$$

D.F. = 1.0, gives;

$$F_r = 87.2 = 3\Delta_s E(I/D.F.)/L^3 + \Sigma \ K\Delta_s$$

$$I_{eq} = \frac{87.2 \times (60 \times 12)^3}{3 \times 20 \times 30 \times 10^3}$$

or: $I_{req} = 18081.8 \text{ in}^4$

Try W36 x 280 $(I = 18,900. \text{ in}^4)$

$$M = 87 \times 60 \times 12 = 62640. \text{ k-in.}$$

$$f = \frac{62640}{1030} = 60 \text{ ksi} \quad (H.S. - \text{steel})$$

However if lateral top support is provided;

$$M_{max} = \frac{3}{16} PL$$

$$f = 60 \times 3/16 = 11.4 \text{ ksi ok.}$$

Bulkhead Effect

for $D_y = 2 \times 10^4$ k-in and D_x k-in and $D_x = 4 \times 10^3$ k-in

then the distribution factor; DF = .13; (Fig. 16)

therefore:

$I_{req} = 18081.8 \times .13$

$I_{req} = 2350.$ in^4

use W24 x 84   (I = 2370 in^4) and   S = 197 in^3

M = 720 x 87 = 62640

$f = \dfrac{62640 \times .13}{197} = 41$ ksi

top support

f = 41 x 3/16 = 7.25 ksi; therefore o.k.

reduction in weight 280 to 84# per foot, or a

saving of 200#/ft along the pile length.

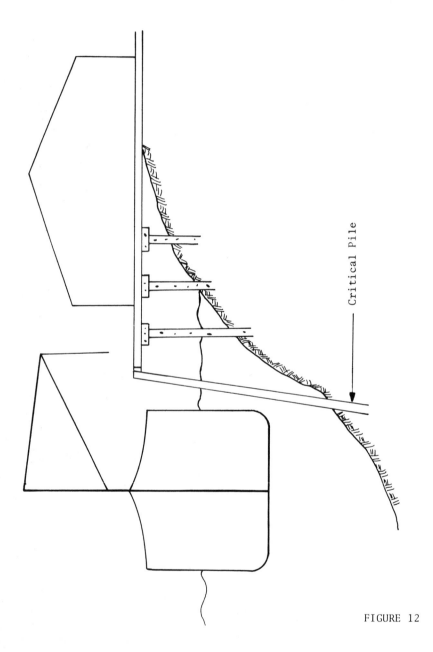

Critical Pile

FIGURE 12

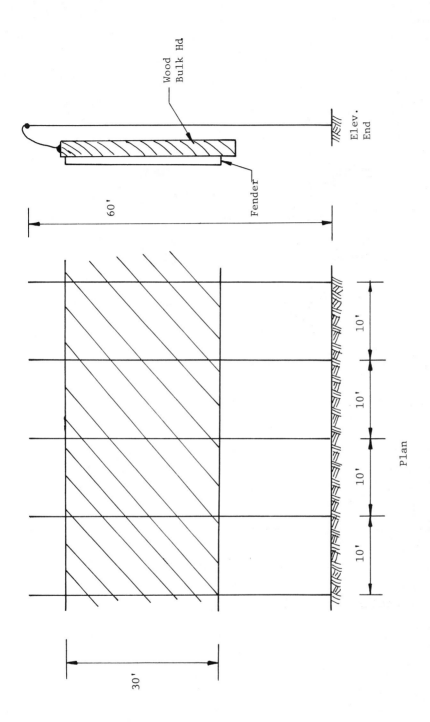

Wood
Bulk Hd

Fender

60'

Elev.
End

30'

10' 10' 10' 10'

Plan

FIGURE 13

155

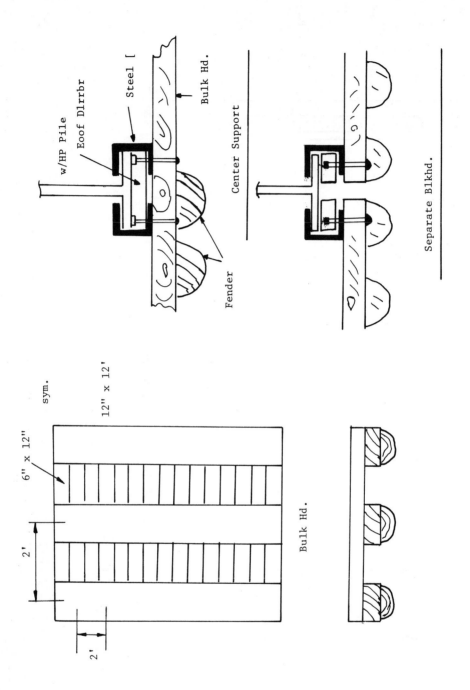

w/HP Pile

Eoof Dlrrbr

Steel [

Bulk Hd.

Fender

Center Support

Separate Blkhd.

sym.

12" x 12'

6" x 12"

2'

2'

Bulk Hd.

FIGURE 14

156

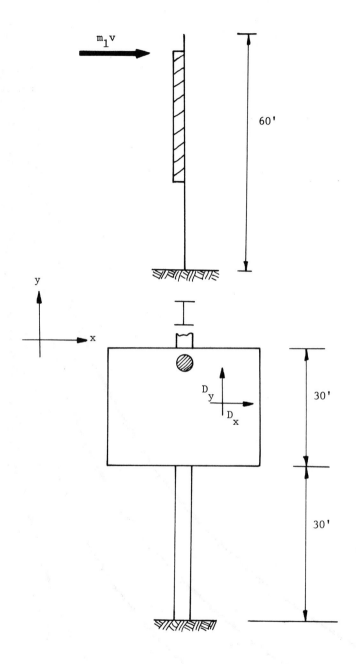

FIGURE 15

157

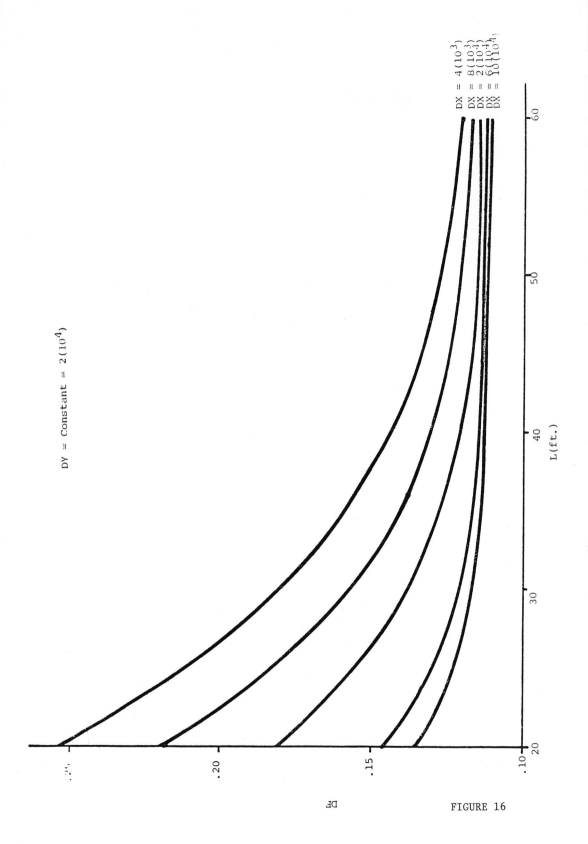

FIGURE 16

EXAMPLE 3

MOORING DOLPHIN

i) Classical Solution

As described previously the classical energy methods can be applied to the design of a mooring dolphin, arranged as shown in Figure 17, with a free cantilever length of 48' and HP 14 x 73 steel sections. The dolphin is impacted with a 25,000 ton vessel at a velocity of 0.50 ft/sec.

The induced dynamic force is given by;

$$F = M_a = M(v_i^2 - v_f^2)/2\Delta$$

where: M = 1,750,000 for 25,000 ship

v_i = 0.5 ft/sec.

Δ = 6 in. (assumed)

The force $F = \dfrac{1.75 \times 10^6 \times (0.5)^2}{2 \times 0.5} = 437.5^K$

Considering a cantilever pile:

$$\Delta = PL^3/3EI$$

or the required stiffness is;

$I_r = PL^3/3E\Delta$ or

$I_r = \dfrac{437.5 \times (48)^3 \times 1728}{3 \times 29 \times 10^3 \times 0.5}$

I_r = 160,167.7 in.^{4}

Assuming a nine pile grouping, as shown in Figure 17, then I_{act} is computed as;

$$I_{act} = 6[I_o + \Delta(36)^2] + 3 \times I_o$$

assuming a HP 14 x 73 section (I_o = 734.in^4, A = 21.5 in^2), gives;

$$I_{act} = 6[734. + 2.15(36)^2] + 3 \times 734$$

$$I_{act} = 173,790 \text{ in}^4 > 160,167.7 \text{ in.}^4$$

Therefore stiffness is satisfied. Assuming the load to each pile is P = 437.5/9 = 48.6, then the moment is M = 48.6 x 48 = 2333.3 k-ft. the induced stress is therefore;

$$f = \frac{2333.3 \times 12}{108} = 259 \text{ ksi}$$

if the group action is assumed then

$$f = \frac{437.5 \times (48 \times 12)}{173790} \times 36 = 52 \text{ ksi}$$

which indicates an over stressed condition.

ii) Computer Solution

The same mooring dolphin has been examined by a dynamic system.

The results of the computer analysis indicate that under impact of the ship, the maxim deformation will be 4.0 in. when subjected to an imapct force of 1721. kips. The bending stress at the mud line for each pile is;

Pile No.	Stress(ksi)	Deflection(in)
1	10.1	4.0
2	0.	0.
3	9.1	3.5
4	5.1	3.6
5	9.1	3.5
6	0.	0.
7	9.1	3.5
8	5.1	3.6
9	9.1	3.5

A summary of the output dynamic data and energy is given in Tables 3 and 4.

A summary of the results shows the system is adequate relative to the strength of the steel sections and soil.

PIER DOLPHIN

i) Classical Solution

The second problem under study consists of five pairs of steel HP 14 x 73 sections (one vertical/one battered) connected by a rigid pier cap, as shown in Fig. 18. The pier will be impacted by a 25,000 Ton vessel at 6 ft/sec.

$$F = M_a = \frac{W}{g} (v^2)/2\Delta$$

the weight of ship is $50,000^K$, and $\Delta = 3$ in, $v_i = 0.5$ ft/sec, which gives;

$$F = \frac{50,000.}{32.2} (\frac{(.5)^2}{2x.25})$$

$$F = 776^K$$

assume a five pile group, therefore;

$$P = 776/5 = 155^K$$

Assuming a batter of 1/2, then the force on the piling is:

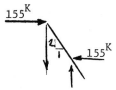

$$V = 2 \times 155 = 310^K$$

Considering a HP x 73 section ($A = 21.46$ in^2), the induced stress is

$$f = 310/21.46 = 14.4 \text{ ksi}$$

ii) Computer solution

Application of the computer program results in a maximum force of 2148.K, inducing a maximum deformation of 1.4 in. and a stopping time of .44 seconds to the system. The resulting stresses for each pile are as follows;

Pile No.	Stress(ksi)	Deflection(in)
1	3.2	1.3
2	3.2	1.3
3	0.2	.67
4	7.0	1.5
5	7.0	1.5
6	7.0	1.5
7	7.0	1.5
8	7.0	1.5
9	2.0	1.2
10	3.2	1.3

The results indicate adequate strength of the steel elements and soil. However the size of the pile could be drastically reduced. A tabulation of the data is given in Tables 5 and 6.

163

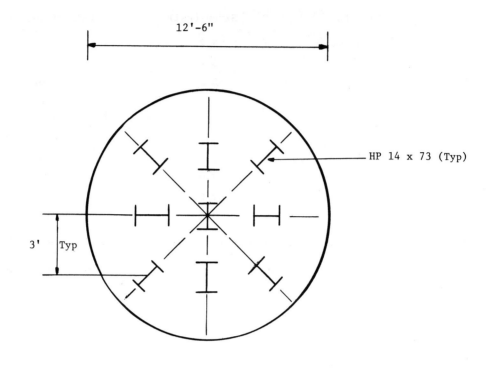

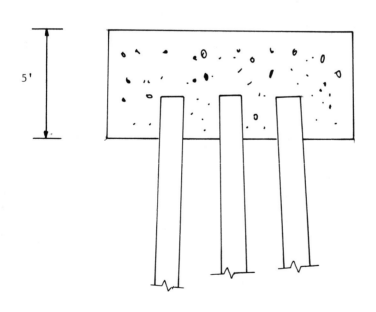

164

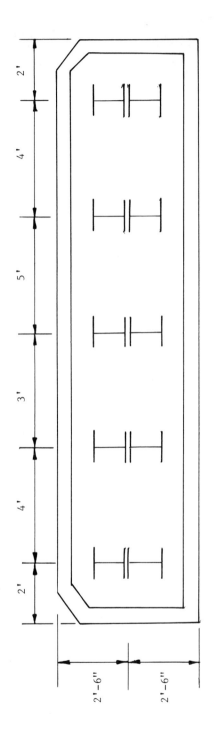

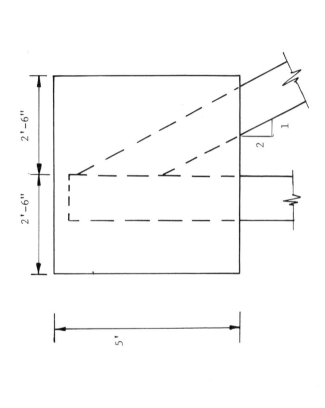

```
                    BASIC PROGRAM PARAMETERS
                    ------------------------

              WEIGHT OF SHIP =     25000.00 TONS  (   56000.00 KIPS)
            VELOCITY OF SHIP =        .32 KNOTS  (     6.48 IPS)
   LENGTH OF DOLPHIN BELOW ML =     65.00 FEET
   LENGTH OF DOLPHIN ABOVE ML =     48.00 FEET
                MASS OF SHIP =       145.05 KIPS/IN/SEC/SEC
       ANGLE OF PASSIVE FAILURE =   45.00 DEGREES
            DEPTH CORRECTION =         .00 FEET
          DISSIPATION FACTOR =        .5000
            ALLOWABLE STRAIN =        .03000 IN/IN
```

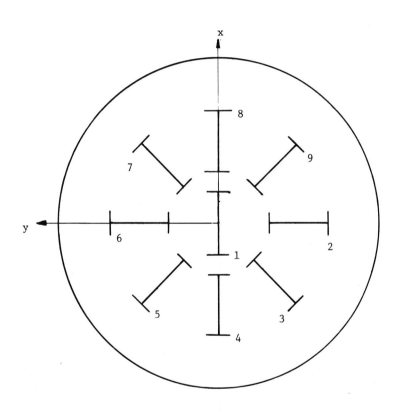

TABLE 3

PILE # 1 /SOIL LOADS AND DEFORMATIONS AT TIME =   .93586 SECONDS

DEPTH BELOW M.L. (IN)	SOIL MODULUS (K/SQ IN)	SOIL RESISTANCE ALLOW. (K/IN)	ACTUAL (K/IN)	PILE 1 DEFL. DEFL (IN)	SLOPE (RAD)	PILE 1 M AND V MOMENT (IN-K)	SHEAR (K)
TOP	.00	.0	.0	4.005	.010	0.	1708.9
.0	.00	.0	.0	.095	.001	1076.	1698.6
39.0	23.40	273.4	1.6	.068	.001	5325.	1644.4
78.0	46.80	546.9	2.0	.043	.001	9575.	1571.2
117.0	70.20	820.3	1.6	.023	.000	12318.	1509.8
156.0	93.60	1093.7	.9	.010	.000	11064.	1474.5
195.0	117.00	1367.1	.2	.002	.000	8431.	1465.6
234.0	140.40	1640.6	-.3	-.002	.000	5452.	1475.7
273.0	163.80	1914.0	-.5	-.003	.000	2866.	1494.8
312.0	187.20	2187.4	-.5	-.003	-.000	1027.	1514.8
351.0	210.60	2460.9	-.4	-.002	-.000	-34.	1530.5
390.0	234.00	2734.3	-.3	-.001	-.000	-481.	1540.5
429.0	257.40	3007.7	-.1	-.000	-.000	-540.	1545.1
468.0	280.80	3281.1	-.0	-.000	-.000	-415.	1546.3
507.0	304.20	3554.6	.0	-.000	-.000	-248.	1545.5
546.0	327.60	3828.0	.0	.000	-.000	-111.	1544.2
585.0	351.00	4101.4	.0	.000	.000	-25.	1543.0
624.0	374.40	4374.9	.0	.000	.000	13.	1542.2
663.0	397.80	4648.3	.0	.000	.000	22.	1541.8
702.0	421.20	4921.7	.0	.000	.000	15.	1541.7
741.0	444.60	5195.2	-.0	-.000	.000	5.	1541.8
780.0	468.00	5468.6	-.0	-.000	.000	0.	1542.0

DOLPHIN/SOIL LOADS AND DEFORMATIONS AT TIME =   .93586 SECONDS

DEPTH BELOW M.L. (IN)	BENDING ALLOWABLE (KSI)	ACTUAL (KSI)	SHEAR ALLOWABLE (KSI)	ACTUAL (KSI)	COMBINED STRESS RATIO
.0	23.80	10.11	14.40	197.91	189.325
39.0	23.80	50.04	14.40	191.60	179.142
78.0	23.80	89.98	14.40	183.07	165.407
117.0	23.80	115.76	14.40	175.92	154.109
156.0	23.80	103.97	14.40	171.80	146.706
195.0	23.80	79.23	14.40	170.77	143.958
234.0	23.80	51.23	14.40	171.94	144.723
273.0	23.80	26.93	14.40	174.17	147.427
312.0	23.80	9.65	14.40	176.50	150.630
351.0	23.80	-.32	14.40	178.33	153.382
390.0	23.80	-4.52	14.40	179.49	155.555
429.0	23.80	-5.07	14.40	180.04	156.527
468.0	23.80	-3.90	14.40	180.17	156.702
507.0	23.80	-2.33	14.40	180.08	156.482
546.0	23.80	-1.04	14.40	179.92	156.156
585.0	23.80	-.24	14.40	179.78	155.881
624.0	23.80	.13	14.40	179.69	155.718
663.0	23.80	.20	14.40	179.65	155.644
702.0	23.80	.14	14.40	179.64	155.625
741.0	23.80	.05	14.40	179.65	155.647
780.0	23.80	.00	14.40	179.67	155.671

PILE # 2 /SOIL LOADS AND DEFORMATIONS AT TIME =   .93586SECONDS

DEPTH BELOW M.L. (IN)	SOIL MODULUS (K/SQ IN)	SOIL RESISTANCE ALLOW. (K/IN)	ACTUAL (K/IN)	PILE 2 DEFL. DEFL (IN)	SLOPE (RAD)	PILE 2 M AND V MOMENT (IN-K)	SHEAR (K)
TOP	.00	.0	.0	.000	.000	0.	1708.9
.0	.00	.0	.0	.000	.000	0.	1708.9
39.0	23.40	273.4	.0	.000	.000	0.	1708.9
78.0	46.80	546.9	.0	.000	.000	0.	1708.9
117.0	70.20	820.3	.0	.000	.000	0.	1708.9
156.0	93.60	1093.7	.0	.000	.000	0.	1708.9
195.0	117.00	1367.1	.0	.000	.000	0.	1708.9
234.0	140.40	1640.6	-.0	-.000	.000	0.	1708.9
273.0	163.80	1914.0	-.0	-.000	.000	0.	1708.9
312.0	187.20	2187.4	-.0	-.000	-.000	0.	1708.9
351.0	210.60	2460.9	-.0	-.000	-.000	-0.	1708.9
390.0	234.00	2734.3	-.0	-.000	-.000	-0.	1708.9
429.0	257.40	3007.7	-.0	-.000	-.000	-0.	1708.9
468.0	280.80	3281.1	-.0	-.000	-.000	-0.	1708.9
507.0	304.20	3554.6	.0	.000	-.000	-0.	1708.9
546.0	327.60	3828.0	.0	.000	.000	-0.	1708.9
585.0	351.00	4101.4	.0	.000	.000	-0.	1708.9
624.0	374.40	4374.9	.0	.000	.000	0.	1708.9
663.0	397.80	4648.3	.0	.000	.000	0.	1708.9
702.0	421.20	4921.7	.0	.000	.000	0.	1708.9
741.0	444.60	5195.2	-.0	-.000	.000	0.	1708.9
780.0	468.00	5468.6	-.0	-.000	.000	0.	1708.9

DOLPHIN/SOIL LOADS AND DEFORMATIONS AT TIME =   .93586 SECONDS

DEPTH BELOW M.L. (IN)	BENDING ALLOWABLE (KSI)	ACTUAL (KSI)	SHEAR ALLOWABLE (KSI)	ACTUAL (KSI)	COMBINED STRESS RATIO
.0	23.80	.00	14.40	199.11	191.193
39.0	23.80	.00	14.40	199.11	191.193
78.0	23.80	.00	14.40	199.11	191.193
117.0	23.80	.00	14.40	199.11	191.193
156.0	23.80	.00	14.40	199.11	191.193
195.0	23.80	.00	14.40	199.11	191.193
234.0	23.80	.00	14.40	199.11	191.193
273.0	23.80	.00	14.40	199.11	191.193
312.0	23.80	.00	14.40	199.11	191.193
351.0	23.80	-.00	14.40	199.11	191.193
390.0	23.80	-.00	14.40	199.11	191.193
429.0	23.80	-.00	14.40	199.11	191.193
468.0	23.80	-.00	14.40	199.11	191.193
507.0	23.80	-.00	14.40	199.11	191.193
546.0	23.80	-.00	14.40	199.11	191.193
585.0	23.80	-.00	14.40	199.11	191.193
624.0	23.80	.00	14.40	199.11	191.193
663.0	23.80	.00	14.40	199.11	191.193
702.0	23.80	.00	14.40	199.11	191.193
741.0	23.80	.00	14.40	199.11	191.193
780.0	23.80	.00	14.40	199.11	191.193

PILE # 3 /SOIL LOADS AND DEFORMATIONS AT TIME = .93586SECONDS

DEPTH BELOW M.L. (IN)	SOIL MODULUS (K/SQ IN)	SOIL RESISTANCE ALLOW. (K/IN)	ACTUAL (K/IN)	PILE 3 DEFL. DEFL (IN)	SLOPE (RAD)	PILE 3 M AND V MOMENT (IN-K)	SHEAR (K)
TOP	.00	.0	.0	-3.509	-.009	0.	1706.9
.0	.00	.0	.0	-.083	-.001	-971.	1717.9
39.0	23.40	273.4	-1.4	-.059	-.001	-4691.	1765.4
78.0	46.80	546.9	-1.8	-.038	-.000	-8411.	1829.6
117.0	70.20	820.3	-1.4	-.020	-.000	-10811.	1883.4
156.0	93.60	1093.7	-.8	-.008	-.000	-9706.	1914.3
195.0	117.00	1367.1	-.2	-.001	-.000	-7394.	1922.1
234.0	140.40	1640.6	.3	.002	-.000	-4780.	1913.2
273.0	163.80	1914.0	.5	.003	-.000	-2511.	1896.4
312.0	187.20	2187.4	.5	.002	.000	-899.	1878.9
351.0	210.60	2460.9	.4	.002	.000	31.	1865.1
390.0	234.00	2734.3	.2	.001	.000	422.	1856.4
429.0	257.40	3007.7	.1	.000	.000	474.	1852.2
468.0	280.80	3281.1	.0	.000	.000	364.	1851.3
507.0	304.20	3554.6	-.0	-.000	.000	218.	1851.9
546.0	327.60	3828.0	-.0	-.000	.000	97.	1853.1
585.0	351.00	4101.4	-.0	-.000	-.000	22.	1854.2
624.0	374.40	4374.9	-.0	-.000	-.000	-12.	1854.9
663.0	397.80	4648.3	-.0	-.000	-.000	-19.	1855.0
702.0	421.20	4921.7	-.0	-.000	-.000	-13.	1855.3
741.0	444.60	5195.2	.0	.000	-.000	-4.	1855.2
780.0	468.00	5468.6	.0	.000	-.000	0.	1855.0

DOLPHIN/SOIL LOADS AND DEFORMATIONS AT TIME = .93586 SECONDS

DEPTH BELOW M.L. (IN)	BENDING ALLOWABLE (KSI)	ACTUAL (KSI)	SHEAR ALLOWABLE (KSI)	ACTUAL (KSI)	COMBINED STRESS RATIO
.0	23.80	-9.13	14.40	200.16	193.598
39.0	23.80	-44.08	14.40	205.70	205.904
78.0	23.80	-79.04	14.40	213.18	222.479
117.0	23.80	-101.59	14.40	219.44	236.502
156.0	23.80	-91.21	14.40	223.05	243.766
195.0	23.80	-69.48	14.40	223.96	244.798
234.0	23.80	-44.92	14.40	222.92	241.539
273.0	23.80	-23.60	14.40	220.96	236.449
312.0	23.80	-8.45	14.40	218.92	231.488
351.0	23.80	.29	14.40	217.31	227.756
390.0	23.80	3.97	14.40	216.30	225.789
429.0	23.80	4.45	14.40	215.82	224.810
468.0	23.80	3.42	14.40	215.71	224.533
507.0	23.80	2.04	14.40	215.78	224.637
546.0	23.80	.91	14.40	215.92	224.875
585.0	23.80	.21	14.40	216.04	225.100
624.0	23.80	-.11	14.40	216.12	225.262
663.0	23.80	-.18	14.40	216.16	225.346
702.0	23.80	-.12	14.40	216.17	225.362
741.0	23.80	-.04	14.40	216.16	225.331
780.0	23.80	.00	14.40	216.14	225.302

PILE # 4 /SOIL LOADS AND DEFORMATIONS AT TIME = .93586 SECONDS

DEPTH BELOW M.L. (IN)	SOIL MODULUS (K/SQ IN)	SOIL RESISTANCE ALLOW. (K/IN)	ACTUAL (K/IN)	PILE 4 DEFL. DEFL (IN)	SLOPE (RAD)	PILE 4 M MOMENT (IN-K)	AND V SHEAR (K)
TOP .0	.00	.0	.0	-3.614	-.009	0.	1708.9
.0	.00	.0	.0	-.084	-.001	-541.	1718.0
39.0	23.40	273.4	-1.4	-.060	+.001	-4421.	1766.3
78.0	46.80	546.9	-1.8	-.038	+.001	-8301.	1831.7
117.0	70.20	820.3	-1.5	-.021	-.000	-10838.	1886.8
156.0	93.60	1093.7	-.8	-.009	-.000	-9792.	1918.9
195.0	117.00	1367.1	-.2	+.002	-.000	-7496.	1927.3
234.0	140.40	1640.6	-.3	.002	-.000	-4570.	1918.8
273.0	163.80	1914.0	.5	.003	-.000	-2577.	1902.0
312.0	187.20	2187.4	.5	.003	.000	-938.	1884.4
351.0	210.60	2460.9	.4	.002	.000	14.	1870.4
390.0	234.00	2734.3	.2	.001	.000	418.	1861.5
429.0	257.40	3007.7	.1	.000	.000	476.	1857.3
468.0	280.80	3281.1	.0	.000	.000	369.	1856.2
507.0	304.20	3554.6	-.0	-.000	.000	222.	1856.9
546.0	327.60	3828.0	-.0	-.000	.000	100.	1858.0
585.0	351.00	4101.4	-.0	-.000	-.000	-23.	1859.1
624.0	374.40	4374.9	-.0	-.000	-.000	-11.	1859.8
663.0	397.80	4648.3	-.0	-.000	-.000	-19.	1860.1
702.0	421.20	4921.7	-.0	-.000	-.000	-13.	1860.2
741.0	444.60	5195.2	.0	.000	-.000	-4.	1860.1
780.0	468.00	5468.6	.0	.000	-.000	0.	1860.0

DOLPHIN/SOIL LOADS AND DEFORMATIONS AT TIME = .93586 SECONDS

DEPTH BELOW M.L. (IN)	BENDING ALLOWABLE (KSI)	ACTUAL (KSI)	SHEAR ALLOWABLE (KSI)	ACTUAL (KSI)	COMBINED STRESS RATIO
.0	23.80	-5.08	14.40	200.18	193.458
39.0	23.80	-41.54	14.40	205.90	206.000
78.0	23.80	-78.00	14.40	213.42	222.941
117.0	23.80	-101.84	14.40	219.85	237.362
156.0	23.80	-92.02	14.40	223.58	244.940
195.0	23.80	-70.44	14.40	224.57	245.166
234.0	23.80	-45.76	14.40	223.57	242.979
273.0	23.80	-24.21	14.40	221.62	237.875
312.0	23.80	-8.81	14.40	219.56	232.857
351.0	23.80	.13	14.40	217.93	229.045
390.0	23.80	3.93	14.40	216.90	227.035
429.0	23.80	4.47	14.40	216.40	226.297
468.0	23.80	3.47	14.40	216.28	225.734
507.0	23.80	2.08	14.40	216.36	225.833
546.0	23.80	.94	14.40	216.49	226.072
585.0	23.80	.22	14.40	216.62	226.299
624.0	23.80	-.11	14.40	216.70	226.465
663.0	23.80	-.18	14.40	216.74	226.551
702.0	23.80	-.12	14.40	216.75	226.567
741.0	23.80	-.04	14.40	216.74	226.536
780.0	23.80	.00	14.40	216.72	226.507

PILE # 5 /SOIL LOADS AND DEFORMATIONS AT TIME = .93586 SECONDS

DEPTH BELOW M.L. (IN)	SOIL MODULUS (K/SQ IN)	SOIL RESISTANCE ALLOW. (K/IN)	ACTUAL (K/IN)	PILE 5 DEFL. DEFL (IN)	SLOPE (RAD)	PILE 5 M AND V MOMENT (IN-K)	SHEAR (K)
TOP	.00	.0	.0	-3.509	-.009	0.	1708.9
.0	.00	.0	.0	-.083	-.001	-971.	1717.9
39.0	23.40	273.4	-1.4	-.059	-.001	-4691.	1765.4
78.0	46.80	546.9	-1.8	-.038	-.000	-8411.	1829.6
117.0	70.20	820.3	-1.4	-.020	-.000	-10811.	1883.4
156.0	93.60	1093.7	-.8	-.008	-.000	-9706.	1914.3
195.0	117.00	1367.1	-.2	-.001	-.000	-7394.	1922.1
234.0	140.40	1640.6	.3	.002	-.000	-4780.	1913.2
273.0	163.80	1914.0	.5	.003	-.000	-2511.	1896.4
312.0	187.20	2187.4	.5	.002	.000	-899.	1878.9
351.0	210.60	2460.9	.4	.002	.000	31.	1865.1
390.0	234.00	2734.3	.2	.001	.000	422.	1856.4
429.0	257.40	3007.7	.1	.000	.000	474.	1852.2
468.0	280.80	3281.1	-.0	.000	.000	364.	1851.3
507.0	304.20	3554.6	-.0	-.000	.000	218.	1851.9
546.0	327.60	3828.0	-.0	-.000	.000	97.	1853.1
585.0	351.00	4101.4	-.0	-.000	-.000	22.	1854.2
624.0	374.40	4374.9	-.0	-.000	-.000	-12.	1854.9
663.0	397.80	4648.3	-.0	-.000	-.000	-19.	1855.2
702.0	421.20	4921.7	-.0	-.000	-.000	-13.	1855.3
741.0	444.60	5195.2	.0	.000	-.000	-4.	1855.2
780.0	468.00	5468.6	.0	.000	-.000	0.	1855.0

DOLPHIN/SOIL LOADS AND DEFORMATIONS AT TIME = .93586 SECONDS

DEPTH BELOW M.L. (IN)	BENDING ALLOWABLE (KSI)	ACTUAL (KSI)	SHEAR ALLOWABLE (KSI)	ACTUAL (KSI)	COMBINED STRESS RATIO
.0	23.80	-9.13	14.40	200.16	193.598
39.0	23.80	-44.08	14.40	205.70	205.904
78.0	23.80	-79.04	14.40	213.18	222.479
117.0	23.80	-101.59	14.40	219.44	236.502
156.0	23.80	-91.21	14.40	223.05	243.766
195.0	23.80	-69.48	14.40	223.96	244.798
234.0	23.80	-44.92	14.40	222.92	241.539
273.0	23.80	-23.60	14.40	220.96	236.449
312.0	23.80	-8.45	14.40	218.92	231.488
351.0	23.80	.29	14.40	217.31	227.756
390.0	23.80	3.97	14.40	216.30	225.789
429.0	23.80	4.45	14.40	215.82	224.810
468.0	23.80	3.42	14.40	215.71	224.533
507.0	23.80	2.04	14.40	215.78	224.637
546.0	23.80	.91	14.40	215.92	224.875
585.0	23.80	.21	14.40	216.04	225.100
624.0	23.80	-.11	14.40	216.12	225.262
663.0	23.80	-.16	14.40	216.16	225.346
702.0	23.80	-.12	14.40	216.17	225.362
741.0	23.80	-.04	14.40	216.16	225.331
780.0	23.80	.00	14.40	216.14	225.302

PILE # 6 /SOIL LOADS AND DEFORMATIONS AT TIME = .93586 SECONDS

DEPTH BELOW M.L. (IN)	SOIL MODULUS (K/SQ IN)	SOIL RESISTANCE ALLOW. (K/IN)	ACTUAL (K/IN)	PILE 6 DEFL. DEFL (IN)	SLOPE (RAD)	PILE 6 M AND V MOMENT (IN-K)	SHEAR (K)
TOP	.00	.0	.0	.000	.000	0.	1708.9
.0	.00	.0	.0	.000	.000	0.	1708.9
39.0	23.40	273.4	.0	.000	.000	0.	1708.9
78.0	46.80	546.9	.0	.000	.000	0.	1708.9
117.0	70.20	820.3	.0	.000	.000	0.	1708.9
156.0	93.60	1093.7	.0	.000	.000	0.	1708.9
195.0	117.00	1367.1	.0	.000	.000	0.	1708.9
234.0	140.40	1640.6	-.0	-.000	.000	0.	1708.9
273.0	163.80	1914.0	-.0	-.000	.000	0.	1708.9
312.0	187.20	2187.4	-.0	-.000	-.000	0.	1708.9
351.0	210.60	2460.9	-.0	-.000	-.000	-0.	1708.9
390.0	234.00	2734.3	-.0	-.000	-.000	-0.	1708.9
429.0	257.40	3007.7	-.0	-.000	-.000	-0.	1708.9
468.0	280.80	3281.1	-.0	-.000	-.000	-0.	1708.9
507.0	304.20	3554.6	.0	.000	-.000	-0.	1708.9
546.0	327.60	3828.0	.0	.000	.000	-0.	1708.9
585.0	351.00	4101.4	.0	.000	.000	-0.	1708.9
624.0	374.40	4374.9	.0	.000	.000	0.	1708.9
663.0	397.80	4648.3	.0	.000	.000	0.	1708.9
702.0	421.20	4921.7	.0	.000	.000	0.	1708.9
741.0	444.60	5195.2	-.0	-.000	.000	0.	1708.9
780.0	468.00	5468.6	-.0	-.000	.000	0.	1708.9

DOLPHIN/SOIL LOADS AND DEFORMATIONS AT TIME = .93586 SECONDS

DEPTH BELOW M.L. (IN)	BENDING ALLOWABLE (KSI)	ACTUAL (KSI)	SHEAR ALLOWABLE (KSI)	ACTUAL (KSI)	COMBINED STRESS RATIO
.0	23.80	.00	14.40	199.11	191.193
39.0	23.80	.00	14.40	199.11	191.193
78.0	23.80	.00	14.40	199.11	191.193
117.0	23.80	.00	14.40	199.11	191.193
156.0	23.80	.00	14.40	199.11	191.193
195.0	23.80	.00	14.40	199.11	191.193
234.0	23.80	.00	14.40	199.11	191.193
273.0	23.80	.00	14.40	199.11	191.193
312.0	23.80	.00	14.40	199.11	191.193
351.0	23.80	-.00	14.40	199.11	191.193
390.0	23.80	-.00	14.40	199.11	191.193
429.0	23.80	-.00	14.40	199.11	191.193
468.0	23.80	-.00	14.40	199.11	191.193
507.0	23.80	-.00	14.40	199.11	191.193
546.0	23.80	-.00	14.40	199.11	191.193
585.0	23.80	.00	14.40	199.11	191.193
624.0	23.80	.00	14.40	199.11	191.193
663.0	23.80	.00	14.40	199.11	191.193
702.0	23.80	.00	14.40	199.11	191.193
741.0	23.80	.00	14.40	199.11	191.193
780.0	23.80	.00	14.40	199.11	191.193

PILE # 7 /SOIL LOADS AND DEFORMATIONS AT TIME = .93586 SECONDS

DEPTH BELOW M.L. (IN)	SOIL MODULUS (K/SQ IN)	SOIL RESISTANCE ALLOW. (K/IN)	ACTUAL (K/IN)	PILE 7 DEFL. DEFL (IN)	SLOPE (RAD)	PILE 7 M AND V MOMENT (IN-K)	SHEAR (K)
TOP	.00	.0	.0	3.509	.009	0.	1708.9
.0	.00	.0	.0	.083	.001	971.	1699.8
39.0	23.40	273.4	1.4	.059	.001	4691.	1652.3
78.0	46.80	546.9	1.8	.038	.000	8411.	1588.2
117.0	70.20	820.3	1.4	.020	.000	10811.	1534.4
156.0	93.60	1093.7	.8	.008	.000	9706.	1503.4
195.0	117.00	1367.1	.2	.001	.000	7394.	1495.7
234.0	140.40	1640.6	-.3	-.002	.000	4780.	1504.5
273.0	163.80	1914.0	-.5	-.003	.000	2511.	1521.3
312.0	187.20	2187.4	-.5	-.002	-.000	899.	1538.8
351.0	210.60	2460.9	-.4	-.001	-.000	-31.	1552.7
390.0	234.00	2734.3	-.2	-.001	-.000	-422.	1561.4
429.0	257.40	3007.7	-.1	-.000	-.000	-474.	1565.5
468.0	280.80	3281.1	-.0	-.000	-.000	-364.	1566.0
507.0	304.20	3554.6	-.0	.000	-.000	-218.	1565.8
546.0	327.60	3828.0	.0	.000	-.000	-97.	1564.6
585.0	351.00	4101.4	.0	.000	.000	-22.	1563.6
624.0	374.40	4374.9	.0	.000	.000	12.	1562.9
663.0	397.80	4648.3	.0	.000	.000	19.	1562.5
702.0	421.20	4921.7	.0	.000	.000	13.	1562.5
741.0	444.60	5195.2	-.0	-.000	.000	4.	1562.6
780.0	468.00	5468.6	-.0	-.000	.000	0.	1562.7

DOLPHIN/SOIL LOADS AND DEFORMATIONS AT TIME = .93586 SECONDS

DEPTH BELOW M.L. (IN)	BENDING ALLOWABLE (KSI)	ACTUAL (KSI)	SHEAR ALLOWABLE (KSI)	ACTUAL (KSI)	COMBINED STRESS RATIO
.0	23.80	9.13	14.40	198.06	189.565
39.0	23.80	44.08	14.40	192.53	180.605
78.0	23.80	79.04	14.40	185.05	168.457
117.0	23.80	101.59	14.40	178.78	158.408
156.0	23.80	91.21	14.40	175.17	151.813
195.0	23.80	69.48	14.40	174.27	149.379
234.0	23.80	44.92	14.40	175.30	150.089
273.0	23.80	23.60	14.40	177.26	152.525
312.0	23.80	8.45	14.40	179.30	155.393
351.0	23.80	-.29	14.40	180.91	157.850
390.0	23.80	-3.97	14.40	181.93	159.779
429.0	23.80	-4.45	14.40	182.41	160.642
468.0	23.80	-3.42	14.40	182.52	160.797
507.0	23.80	-2.04	14.40	182.44	160.601
546.0	23.80	-.91	14.40	182.30	160.313
585.0	23.80	-.21	14.40	182.18	160.068
624.0	23.80	.11	14.40	182.10	159.924
663.0	23.80	.18	14.40	182.06	159.855
702.0	23.80	.12	14.40	182.05	159.841
741.0	23.80	.04	14.40	182.07	159.861
780.0	23.80	.00	14.40	182.08	159.882

PILE # 8 /SOIL LOADS AND DEFORMATIONS AT TIME =   .93586SECONDS

DEPTH BELOW M.L. (IN)	SOIL MODULUS (K/SQ IN)	SOIL RESISTANCE ALLOW. (K/IN)	SOIL RESISTANCE ACTUAL (K/IN)	PILE 8 DEFL. DEFL (IN)	PILE 8 DEFL. SLOPE (RAD)	PILE 8 M AND V MOMENT (IN-K)	PILE 8 M AND V SHEAR (K)
TOP	.00	.0	.0	3.614	.009	0.	1708.9
.0	.00	.0	.0	.084	.001	541.	1699.7
39.0	23.40	273.4	1.4	.060	.001	4421.	1651.5
78.0	46.80	546.9	1.8	.038	.001	8301.	1586.0
117.0	70.20	820.3	1.5	.021	.000	10838.	1530.9
156.0	93.60	1093.7	.8	.009	.000	9792.	1498.9
195.0	117.00	1367.1	.2	.002	.000	7496.	1490.4
234.0	140.40	1640.6	-.3	-.002	.000	4870.	1498.9
273.0	163.80	1914.0	-.5	-.003	.000	2577.	1515.7
312.0	187.20	2187.4	-.5	-.003	-.000	938.	1533.3
351.0	210.60	2460.9	-.4	-.002	-.000	-14.	1547.4
390.0	234.00	2734.3	-.2	-.001	-.000	-418.	1556.2
429.0	257.40	3007.7	-.1	-.000	-.000	-476.	1560.5
468.0	280.80	3281.1	-.0	-.000	-.000	-369.	1561.5
507.0	304.20	3554.6	.0	.000	-.000	-222.	1560.9
546.0	327.60	3828.0	.0	.000	-.000	-100.	1559.7
585.0	351.00	4101.4	.0	.000	.000	-23.	1558.6
624.0	374.40	4374.9	.0	.000	.000	11.	1557.9
663.0	397.80	4648.3	.0	.000	.000	19.	1557.5
702.0	421.20	4921.7	.0	.000	.000	13.	1557.5
741.0	444.60	5195.2	-.0	-.000	.000	4.	1557.6
780.0	468.00	5468.6	-.0	-.000	.000	0.	1557.7

DOLPHIN/SOIL LOADS AND DEFORMATIONS AT TIME =   .93586 SECONDS

DEPTH BELOW M.L. (IN)	BENDING ALLOWABLE (KSI)	BENDING ACTUAL (KSI)	SHEAR ALLOWABLE (KSI)	SHEAR ACTUAL (KSI)	COMBINED STRESS RATIO
.0	23.80	5.08	14.40	198.05	189.365
39.0	23.80	41.54	14.40	192.42	180.308
78.0	23.80	78.00	14.40	184.80	167.975
117.0	23.80	101.84	14.40	178.38	157.728
156.0	23.80	92.02	14.40	174.64	150.953
195.0	23.80	70.44	14.40	173.66	148.390
234.0	23.80	45.76	14.40	174.65	149.024
273.0	23.80	24.21	14.40	176.61	151.431
312.0	23.80	8.81	14.40	178.66	154.303
351.0	23.80	-.13	14.40	180.29	156.767
390.0	23.80	-3.93	14.40	181.33	158.731
429.0	23.80	-4.47	14.40	181.82	159.618
468.0	23.80	-3.47	14.40	181.94	159.786
507.0	23.80	-2.08	14.40	181.87	159.596
546.0	23.80	-.94	14.40	181.73	159.307
585.0	23.80	.22	14.40	181.61	159.061
624.0	23.80	.11	14.40	181.53	158.913
663.0	23.80	.18	14.40	181.49	158.846
702.0	23.80	.12	14.40	181.48	158.829
741.0	23.80	.04	14.40	181.49	158.848
780.0	23.80	.00	14.40	181.50	158.869

PILE # 9 /SOIL LOADS AND DEFORMATIONS AT TIME = .93586 SECONDS

DEPTH BELOW M.L. (IN)	SOIL MODULUS (K/SQ IN)	SOIL RESISTANCE ALLOW. (K/IN)	SOIL RESISTANCE ACTUAL (K/IN)	PILE 9 DEFL. DEFL (IN)	PILE 9 DEFL. SLOPE (RAD)	PILE 9 M MOMENT (IN-K)	PILE 9 V SHEAR (K)
TOP	.00	.0	.0	3.509	.009	0.	1708.9
.0	.00	.0	.0	.083	.001	971.	1699.8
39.0	23.40	273.4	1.4	.059	.001	4691.	1652.3
78.0	46.80	546.9	1.8	.038	.000	8411.	1588.2
117.0	70.20	820.3	1.4	.020	.000	10811.	1534.4
156.0	93.60	1093.7	.8	.008	.000	9706.	1503.0
195.0	117.00	1367.1	.2	.001	.000	7394.	1495.7
234.0	140.40	1640.6	-.3	-.002	.000	4780.	1504.5
273.0	163.80	1914.0	-.5	-.003	.000	2511.	1521.3
312.0	187.20	2187.4	-.5	-.002	-.000	899.	1538.8
351.0	210.60	2460.9	-.4	-.002	-.000	-31.	1552.7
390.0	234.00	2734.3	-.2	-.001	-.000	-422.	1561.4
429.0	257.40	3007.7	-.1	-.000	-.000	-474.	1565.5
468.0	280.80	3281.1	-.0	-.000	-.000	-364.	1566.4
507.0	304.20	3554.6	.0	.000	-.000	-219.	1565.8
546.0	327.60	3828.0	.0	.000	-.000	-97.	1564.6
585.0	351.00	4101.4	.0	.000	.000	-22.	1563.6
624.0	374.40	4374.9	.0	.000	.000	12.	1562.9
663.0	397.80	4648.3	.0	.000	.000	19.	1562.5
702.0	421.20	4921.7	.0	.000	.000	13.	1562.5
741.0	444.60	5195.2	-.0	-.000	.000	4.	1562.6
780.0	468.00	5468.6	-.0	-.000	.000	0.	1562.7

DOLPHIN/SOIL LOADS AND DEFORMATIONS AT TIME = .93586 SECONDS

DEPTH BELOW M.L. (IN)	BENDING ALLOWABLE (KSI)	BENDING ACTUAL (KSI)	SHEAR ALLOWABLE (KSI)	SHEAR ACTUAL (KSI)	COMBINED STRESS RATIO
.0	23.80	9.13	14.40	198.06	189.565
39.0	23.80	44.08	14.40	192.53	180.605
78.0	23.80	79.04	14.40	185.05	168.457
117.0	23.80	101.59	14.40	178.78	158.408
156.0	23.80	91.21	14.40	175.17	151.813
195.0	23.80	69.48	14.40	174.27	149.379
234.0	23.80	44.92	14.40	175.30	150.089
273.0	23.80	23.60	14.40	177.26	152.525
312.0	23.80	5.45	14.40	179.30	155.393
351.0	23.80	-.29	14.40	180.91	157.850
390.0	23.80	-3.97	14.40	181.93	159.779
429.0	23.80	-4.45	14.40	182.41	160.642
468.0	23.80	-3.42	14.40	182.52	160.797
507.0	23.80	-2.04	14.40	182.44	160.601
546.0	23.80	-.91	14.40	182.30	160.313
585.0	23.80	-.21	14.40	182.18	160.068
624.0	23.80	.11	14.40	182.10	159.924
663.0	23.80	.18	14.40	182.06	159.858
702.0	23.80	.12	14.40	182.05	159.841
741.0	23.80	.04	14.40	182.07	159.861
780.0	23.80	.00	14.40	182.08	159.882

TABLE 4

OUTPUT OF DYNAMIC DATA AT THE POINT OF IMPACT

STEP NO	TIME (SEC)	DEFL (IN)	VELOCITY (IN/SEC)	ACCELERATION (IN/SEC/SEC)	FORCE (K)	SPRING K (K/IN)
1	.0000	.00	6.48	.00	1.00	1046.11
2	.1170	.75	6.33	-2.53	366.46	487.04
3	.2340	1.47	5.90	-4.94	716.20	487.04
4	.3509	2.12	5.19	-7.12	1033.29	487.04
5	.4679	2.68	4.25	-8.98	1303.26	487.04
6	.5849	3.11	3.11	-10.44	1513.80	487.04
7	.7019	3.40	1.84	-11.41	1655.30	487.04
8	.8189	3.53	-.47	-11.87	1721.33	487.04
9	.9359	3.51	-.91	-11.73	1708.86	487.04

OUTPUT OF KINETIC AND POTENTIAL ENERGY FOR SYSTEM

STEP NO	TIME (SEC.)	KINETIC ENERGY (IN-K)	POTENTIAL ENERGY (IN-K)	TOTAL PE + KE (IN--K)
1	.0000	3046.	0.	3046.
2	.1170	2909.	138.	3047.
3	.2340	2522.	527.	3049.
4	.3509	1955.	1096.	3051.
5	.4679	1309.	1744.	3054.
6	.5849	703.	2353.	3056.
7	.7019	244.	2813.	3058.
8	.8189	16.	3042.	3058.

```
         TOTAL KINETIC ENERGY OF SHIP     =    3046. K-IN
  TOTAL POTENTIAL ENERGY OF DOLPHIN       =    3042. K-IN
                             ERROR        =    .14 PERCENT
```

```
**********************************************************
*****                                                *****
***** MAXIMUM DEFLECTION AT ML =          .084 INCHES *****
*****                                                *****
**********************************************************
```

FACTOR OF SAFETY =   180.016

BASIC PROGRAM PARAMETERS

```
           WEIGHT OF SHIP =    12500.00 TONS  (  28000.00 KIPS)
         VELOCITY OF SHIP =        .32 KNOTS  (      6.48 IPS)
LENGTH OF DOLPHIN BELOW ML =      60.00 FEET
LENGTH OF DOLPHIN ABOVE ML =      30.00 FEET
             MASS OF SHIP =       72.53 KIPS/IN/SEC/SEC
  ANGLE OF PASSIVE FAILURE =      45.00 DEGREES
         DEPTH CORRECTION =         .00 FEET
       DISSIPATION FACTOR =       .5000
         ALLOWABLE STRAIN =       .03000 IN/IN
```

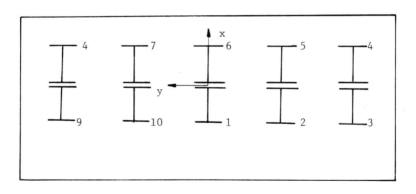

TABLE 5

PILE # 1 /SOIL LOADS AND DEFORMATIONS AT TIME = .43897SECONDS

DEPTH BELOW M.L. (IN)	SOIL MODULUS (K/SQ IN)	SOIL RESISTANCE		PILE 1 DEFL.		PILE 1 M AND V	
		ALLOW. (K/IN)	ACTUAL (K/IN)	DEFL (IN)	SLOPE (RAD)	MOMENT (IN-K)	SHEAR (K)
TOP	.00	.0	.0	1.327	.005	0.	1984.5
.0	.00	.0	.0	.102	.001	338.	1974.8
36.0	21.60	233.0	1.6	.075	.001	5007.	1922.6
72.0	43.20	465.9	2.2	.050	.001	9677.	1849.3
108.0	64.80	698.9	1.9	.030	.000	12992.	1782.5
144.0	86.40	931.8	1.3	.015	.000	12417.	1737.4
180.0	108.00	1164.8	.5	.005	.000	10230.	1718.4
216.0	129.60	1397.7	-.1	-.001	.000	7349.	1720.3
252.0	151.20	1630.7	-.5	-.003	.000	4535.	1735.2
288.0	172.80	1863.6	-.6	-.003	-.000	2259.	1755.2
324.0	194.40	2096.6	-.5	-.003	-.000	703.	1774.0
360.0	216.00	2329.6	-.4	-.002	-.000	-176.	1788.5
396.0	237.60	2562.5	-.2	-.001	-.000	-535.	1797.5
432.0	259.20	2795.5	-.1	-.000	-.000	-569.	1801.9
468.0	280.80	3028.4	-.0	-.000	-.000	-446.	1803.0
504.0	302.40	3261.4	.0	-.000	-.000	-281.	1802.3
540.0	324.00	3494.3	.0	.000	-.000	-141.	1801.0
576.0	345.60	3727.3	.0	.000	.000	-48.	1799.7
612.0	367.20	3960.3	.0	.000	.000	-2.	1798.8
648.0	388.80	4193.2	.0	.000	.000	10.	1798.3
684.0	410.40	4426.2	.0	.000	.000	5.	1798.3
720.0	432.00	4659.1	-.0	-.000	.000	0.	1798.4

DOLPHIN/SOIL LOADS AND DEFORMATIONS AT TIME = .43897 SECONDS

DEPTH BELOW M.L. (IN)	BENDING		SHEAR		COMBINED STRESS RATIO
	ALLOWABLE (KSI)	ACTUAL (KSI)	ALLOWABLE (KSI)	ACTUAL (KSI)	
.0	23.80	3.18	14.40	230.10	255.465
36.0	23.80	47.06	14.40	224.04	244.030
72.0	23.80	90.93	14.40	215.47	227.718
108.0	23.80	122.09	14.40	207.69	213.150
144.0	23.80	116.69	14.40	202.47	202.599
180.0	23.80	96.13	14.40	200.22	197.373
216.0	23.80	69.06	14.40	200.44	196.655
252.0	23.80	42.61	14.40	202.19	198.931
288.0	23.80	21.23	14.40	204.51	202.594
324.0	23.80	6.60	14.40	206.71	206.335
360.0	23.80	-1.65	14.40	208.39	209.486
396.0	23.80	-5.03	14.40	209.44	211.752
432.0	23.80	-5.35	14.40	209.95	212.797
468.0	23.80	-4.19	14.40	210.08	213.014
504.0	23.80	-2.64	14.40	210.00	212.793
540.0	23.80	-1.33	14.40	209.85	212.425
576.0	23.80	-.46	14.40	209.70	212.084
612.0	23.80	-.02	14.40	209.59	211.846
648.0	23.80	.10	14.40	209.53	211.736
684.0	23.80	.05	14.40	209.53	211.730
720.0	23.80	.00	14.40	209.55	211.763

PILE # 2 /SOIL LOADS AND DEFORMATIONS AT TIME =   .43897SECONDS

DEPTH BELOW M.L. (IN)	SOIL MODULUS (K/SQ IN)	SOIL RESISTANCE		PILE 2 DEFL.		PILE 2 M AND V	
		ALLOW. (K/IN)	ACTUAL (K/IN)	DEFL (IN)	SLOPE (RAD)	MOMENT (IN-K)	SHEAR (K)
TOP	.00	.0	.0	1.327	.005	0.	1984.5
.0	.00	.0	.0	.102	.001	337.	1974.8
36.0	21.60	233.0	1.6	.075	.001	5006.	1922.8
72.0	43.20	465.9	2.2	.050	.001	9674.	1849.3
108.0	64.80	698.9	1.9	.030	.000	12989.	1782.5
144.0	86.40	931.8	1.3	.015	.000	12415.	1737.7
180.0	108.00	1164.8	.5	.005	.000	10228.	1718.5
216.0	129.60	1397.7	-.1	-.001	.000	7347.	1720.3
252.0	151.20	1630.7	-.5	-.003	.000	4534.	1735.3
288.0	172.80	1863.6	-.6	-.003	-.000	2259.	1755.2
324.0	194.40	2096.6	-.5	-.003	-.000	702.	1774.1
360.0	216.00	2329.6	-.4	-.002	-.000	-176.	1788.5
396.0	237.60	2562.5	-.2	-.001	-.000	-535.	1797.5
432.0	259.20	2795.5	-.1	-.000	-.000	-569.	1801.9
468.0	280.80	3028.4	-.0	-.000	-.000	-445.	1803.0
504.0	302.40	3261.4	.0	-.000	-.000	-281.	1802.4
540.0	324.00	3494.3	.0	.000	-.000	-141.	1801.0
576.0	345.60	3727.3	.0	.000	.000	-48.	1799.8
612.0	367.20	3960.3	.0	.000	.000	-2.	1798.8
648.0	388.80	4193.2	.0	.000	.000	10.	1798.3
684.0	410.40	4426.2	.0	.000	.000	5.	1798.3
720.0	432.00	4659.1	-.0	-.000	.000	0.	1798.5

DOLPHIN/SOIL LOADS AND DEFORMATIONS AT TIME =   .43897 SECONDS

DEPTH BELOW M.L. (IN)	BENDING		SHEAR		COMBINED STRESS RATIO
	ALLOWABLE (KSI)	ACTUAL (KSI)	ALLOWABLE (KSI)	ACTUAL (KSI)	
.0	23.80	3.17	14.40	230.10	255.465
36.0	23.80	47.04	14.40	224.04	244.032
72.0	23.80	90.91	14.40	215.47	227.724
108.0	23.80	122.06	14.40	207.69	213.159
144.0	23.80	116.66	14.40	202.48	202.609
180.0	23.80	96.11	14.40	200.23	197.384
216.0	23.80	69.04	14.40	200.45	196.666
252.0	23.80	42.60	14.40	202.19	198.942
288.0	23.80	21.23	14.40	204.52	202.603
324.0	23.80	6.60	14.40	206.71	206.344
360.0	23.80	-1.65	14.40	208.39	209.494
396.0	23.80	-5.03	14.40	209.44	211.760
432.0	23.80	-5.35	14.40	209.95	212.805
468.0	23.80	-4.19	14.40	210.09	213.022
504.0	23.80	-2.64	14.40	210.01	212.800
540.0	23.80	-1.32	14.40	209.85	212.432
576.0	23.80	-.46	14.40	209.70	212.091
612.0	23.80	-.02	14.40	209.59	211.853
648.0	23.80	.10	14.40	209.54	211.743
684.0	23.80	.05	14.40	209.54	211.738
720.0	23.80	.00	14.40	209.55	211.771

PILE # 3 /SOIL LOADS AND DEFORMATIONS AT TIME = .43897 SECONDS

DEPTH BELOW M.L. (IN)	SOIL MODULUS (K/SQ IN)	SOIL RESISTANCE ALLOW. (K/IN)	ACTUAL (K/IN)	PILE 3 DEFL. DEFL (IN)	SLOPE (RAD)	PILE 3 M MOMENT (IN-K)	AND V SHEAR (K)
TOP	.00	.0	.0	.672	.002	0.	1984.5
.0	.00	.0	.0	.051	.000	21.	1979.6
36.0	21.60	233.0	.8	.038	.000	2405.	1953.4
72.0	43.20	465.9	1.1	.025	.000	4789.	1916.3
108.0	64.80	698.9	1.0	.015	.000	6490.	1882.6
144.0	86.40	931.8	.6	.007	.000	6226.	1859.8
180.0	108.00	1164.8	.3	.002	.000	5143.	1849.9
216.0	129.60	1397.7	-.0	-.000	.000	3704.	1850.7
252.0	151.20	1630.7	-.2	-.001	.000	2292.	1858.1
288.0	172.80	1863.6	-.3	-.002	-.000	1147.	1868.1
324.0	194.40	2096.6	-.3	-.001	-.000	362.	1877.6
360.0	216.00	2329.6	-.2	-.001	-.000	-83.	1884.8
396.0	237.60	2562.5	-.1	-.001	-.000	-266.	1889.4
432.0	259.20	2795.5	-.1	-.000	-.000	-285.	1891.6
468.0	280.80	3028.4	-.0	-.000	-.000	-224.	1892.4
504.0	302.40	3261.4	.0	.000	-.000	-142.	1891.9
540.0	324.00	3494.3	.0	.000	.000	-71.	1891.2
576.0	345.60	3727.3	.0	.000	.000	-25.	1890.6
612.0	367.20	3960.3	.0	.000	.000	-1.	1890.1
648.0	388.80	4193.2	.0	.000	.000	5.	1889.8
684.0	410.40	4426.2	.0	.000	.000	3.	1889.8
720.0	432.00	4659.1	-.0	-.000	.000	0.	1889.9

DOLPHIN/SOIL LOADS AND DEFORMATIONS AT TIME = .43897 SECONDS

DEPTH BELOW M.L. (IN)	BENDING ALLOWABLE (KSI)	ACTUAL (KSI)	SHEAR ALLOWABLE (KSI)	ACTUAL (KSI)	COMBINED STRESS RATIO
.0	23.80	.20	14.40	230.66	256.593
36.0	23.80	22.60	14.40	227.61	250.785
72.0	23.80	45.00	14.40	223.29	242.329
108.0	23.80	60.99	14.40	219.35	234.601
144.0	23.80	58.51	14.40	216.70	228.922
180.0	23.80	48.33	14.40	215.55	226.088
216.0	23.80	34.80	14.40	215.64	225.707
252.0	23.80	21.54	14.40	216.50	226.951
288.0	23.80	10.78	14.40	217.66	228.935
324.0	23.80	3.40	14.40	218.77	230.946
360.0	23.80	-.78	14.40	219.61	232.623
396.0	23.80	-2.50	14.40	220.15	233.825
432.0	23.80	-2.68	14.40	220.40	234.383
468.0	23.80	-2.11	14.40	220.47	234.503
504.0	23.80	-1.33	14.40	220.43	234.390
540.0	23.80	-.67	14.40	220.36	234.199
576.0	23.80	-.23	14.40	220.28	234.020
612.0	23.80	-.01	14.40	220.23	233.894
648.0	23.80	.05	14.40	220.20	233.835
684.0	23.80	.03	14.40	220.20	233.832
720.0	23.80	.00	14.40	220.21	233.849

PILE # 4 /SOIL LOADS AND DEFORMATIONS AT TIME =   .43897SECONDS

DEPTH BELOW M.L. (IN)	SOIL MODULUS (K/SQ IN)	SOIL RESISTANCE ALLOW. (K/IN)	SOIL RESISTANCE ACTUAL (K/IN)	PILE 4 DEFL. DEFL (IN)	PILE 4 DEFL. SLOPE (RAD)	PILE 4 M AND V MOMENT (IN-K)	PILE 4 M AND V SHEAR (K)
TOP	.00	.0	.0	1.506	.005	0.	1984.5
.0	.00	.0	.0	.117	.001	748.	1973.4
36.0	21.60	233.0	1.9	.086	.001	5999.	1914.0
72.0	43.20	465.9	2.5	.057	.001	11250.	1830.3
108.0	64.80	698.9	2.2	.034	.001	14957.	1754.4
144.0	86.40	931.8	1.4	.016	.000	14240.	1703.7
180.0	108.00	1164.8	.6	.005	.000	11698.	1682.2
216.0	129.60	1397.7	-.1	-.001	.000	8382.	1684.8
252.0	151.20	1630.7	-.5	-.003	.000	5156.	1702.1
288.0	172.80	1863.6	-.7	-.004	-.000	2557.	1725.1
324.0	194.40	2096.6	-.6	-.003	-.000	783.	1746.7
360.0	216.00	2329.6	-.5	-.002	-.000	-214.	1763.1
396.0	237.60	2562.5	-.3	-.001	-.000	-619.	1773.4
432.0	259.20	2795.5	-.1	-.000	-.000	-653.	1778.3
468.0	280.80	3028.4	-.0	-.000	-.000	-509.	1779.6
504.0	302.40	3261.4	.0	-.000	-.000	-321.	1778.8
540.0	324.00	3494.3	.0	.000	-.000	-160.	1777.3
576.0	345.60	3727.3	.0	.000	.000	-54.	1775.8
612.0	367.20	3960.3	.0	.000	.000	-2.	1774.7
648.0	386.80	4193.2	.0	.000	.000	12.	1774.2
684.0	410.40	4426.2	.0	.000	.000	6.	1774.2
720.0	432.00	4659.1	-.0	-.000	.000	6.	1774.3

DOLPHIN/SOIL LOADS AND DEFORMATIONS AT TIME =   .43897 SECONDS

DEPTH BELOW M.L. (IN)	BENDING ALLOWABLE (KSI)	BENDING ACTUAL (KSI)	SHEAR ALLOWABLE (KSI)	SHEAR ACTUAL (KSI)	COMBINED STRESS RATIO
.0	23.80	7.03	14.40	229.94	255.269
36.0	23.80	56.38	14.40	223.02	242.226
72.0	23.80	105.72	14.40	213.26	223.769
108.0	23.80	140.55	14.40	204.42	207.427
144.0	23.80	133.81	14.40	198.52	195.672
180.0	23.80	109.93	14.40	196.01	189.896
216.0	23.80	78.76	14.40	196.30	189.145
252.0	23.80	48.46	14.40	198.33	191.725
288.0	23.80	24.03	14.40	201.00	195.847
324.0	23.80	7.36	14.40	203.52	200.053
360.0	23.80	-2.01	14.40	205.43	203.606
396.0	23.80	-5.81	14.40	206.63	206.148
432.0	23.80	-6.14	14.40	207.21	207.313
468.0	23.80	-4.79	14.40	207.35	207.547
504.0	23.80	-3.01	14.40	207.26	207.290
540.0	23.80	-1.50	14.40	207.08	206.871
576.0	23.80	-.51	14.40	206.91	206.485
612.0	23.80	-.02	14.40	206.79	206.217
648.0	23.80	.11	14.40	206.72	206.094
684.0	23.80	.06	14.40	206.72	206.088
720.0	23.80	.00	14.40	206.74	206.126

183

PILE # 5 /SOIL LOADS AND DEFORMATIONS AT TIME = .43897SECONDS

DEPTH BELOW M.L. (IN)	SOIL MODULUS (K/SQ IN)	SOIL RESISTANCE ALLOW. (K/IN)	ACTUAL (K/IN)	PILE 5 DEFL. DEFL (IN)	SLOPE (RAD)	PILE 5 M AND V MOMENT (IN-K)	SHEAR (K)
TOP .0	.00	.0	.0	1.506	.005	0.	1984.5
.0	.00	.0	.0	.117	.001	749.	1973.4
36.0	21.60	233.0	1.9	.086	.001	6001.	1914.0
72.0	43.20	465.9	2.5	.057	.001	11253.	1830.2
108.0	64.80	698.9	2.2	.034	.001	14961.	1754.4
144.0	86.40	931.8	1.4	.016	.000	14243.	1703.7
180.0	108.00	1164.8	.6	.005	.000	11701.	1682.2
216.0	129.60	1397.7	-.1	-.001	.000	8383.	1684.7
252.0	151.20	1630.7	-.5	-.003	.000	5157.	1702.1
288.0	172.80	1863.6	-.7	-.004	-.000	2557.	1725.0
324.0	194.40	2096.6	-.6	-.003	-.000	783.	1746.6
360.0	216.00	2329.6	-.5	-.002	-.000	-214.	1763.1
396.0	237.60	2562.5	-.3	-.001	-.000	-619.	1773.3
432.0	259.20	2795.5	-.1	-.000	-.000	-653.	1778.3
468.0	280.80	3028.4	-.0	-.000	-.000	-509.	1779.6
504.0	302.40	3261.4	.0	-.000	-.000	-321.	1778.8
540.0	324.00	3494.3	.0	.000	-.000	-160.	1777.2
576.0	345.60	3727.3	.0	.000	.000	-54.	1775.8
612.0	367.20	3960.3	.0	.000	.000	-2.	1774.7
648.0	388.80	4193.2	.0	.000	.000	12.	1774.1
684.0	410.40	4426.2	.0	.000	.000	6.	1774.1
720.0	432.00	4659.1	-.0	-.000	.000	0.	1774.3

DOLPHIN/SOIL LOADS AND DEFORMATIONS AT TIME = .43897 SECONDS

DEPTH BELOW M.L. (IN)	BENDING ALLOWABLE (KSI)	ACTUAL (KSI)	SHEAR ALLOWABLE (KSI)	ACTUAL (KSI)	COMBINED STRESS RATIO
.0	23.80	7.04	14.40	229.94	255.269
36.0	23.80	56.40	14.40	223.02	242.224
72.0	23.80	105.75	14.40	213.26	223.763
108.0	23.80	140.59	14.40	204.41	207.418
144.0	23.80	133.84	14.40	198.51	195.661
180.0	23.80	109.95	14.40	196.00	189.885
216.0	23.80	78.78	14.40	196.30	189.133
252.0	23.80	48.46	14.40	198.32	191.714
288.0	23.80	24.03	14.40	201.00	195.837
324.0	23.80	7.36	14.40	203.51	200.043
360.0	23.80	-2.01	14.40	205.43	203.597
396.0	23.80	-5.82	14.40	206.63	206.139
432.0	23.80	-6.14	14.40	207.20	207.305
468.0	23.80	-4.79	14.40	207.35	207.539
504.0	23.80	-3.01	14.40	207.26	207.282
540.0	23.80	-1.50	14.40	207.08	206.863
576.0	23.80	-.51	14.40	206.91	206.477
612.0	23.80	-.02	14.40	206.78	206.208
648.0	23.80	.11	14.40	206.72	206.086
684.0	23.80	.06	14.40	206.72	206.080
720.0	23.80	.00	14.40	206.74	206.118

PILE # 6 /SOIL LOADS AND DEFORMATIONS AT TIME = .43897 SECONDS

DEPTH BELOW M.L. (IN)	SOIL MODULUS (K/SQ IN)	SOIL RESISTANCE ALLOW. (K/IN)	ACTUAL (K/IN)	PILE 6 DEFL. DEFL (IN)	SLOPE (RAD)	PILE 6 M MOMENT (IN-K)	AND V SHEAR (K)
TOP	.00	.0	.0	1.507	.005	0.	1984.5
.0	.00	.0	.0	.117	.001	751.	1973.4
36.0	21.60	233.0	1.9	.086	.001	6003.	1914.0
72.0	43.20	465.9	2.5	.057	.001	11256.	1830.2
108.0	64.80	698.9	2.2	.034	.001	14964.	1754.3
144.0	86.40	931.8	1.4	.016	.000	14246.	1703.6
180.0	108.00	1164.8	.6	.005	.000	11703.	1682.1
216.0	129.60	1397.7	-.1	-.001	.000	8385.	1684.6
252.0	151.20	1630.7	-.5	-.003	.000	5158.	1702.0
288.0	172.80	1863.6	-.7	-.004	-.000	2558.	1725.0
324.0	194.40	2096.6	-.6	-.003	-.000	783.	1746.6
360.0	216.00	2329.6	-.5	-.002	-.000	-214.	1763.0
396.0	237.60	2562.5	-.3	-.001	-.000	-619.	1773.3
432.0	259.20	2795.5	-.1	-.000	-.000	-653.	1778.3
468.0	280.80	3028.4	-.0	-.000	-.000	-510.	1779.5
504.0	302.40	3261.4	.0	.000	-.000	-321.	1778.7
540.0	324.00	3494.3	.0	.000	-.000	-160.	1777.2
576.0	345.60	3727.3	.0	.000	.000	-54.	1775.7
612.0	367.20	3960.3	.0	.000	.000	-2.	1774.7
648.0	388.80	4193.2	.0	.000	.000	12.	1774.1
684.0	410.40	4426.2	.0	-.000	.000	6.	1774.1
720.0	432.00	4659.1	-.0	-.000	.000	0.	1774.3

DOLPHIN/SOIL LOADS AND DEFORMATIONS AT TIME = .43897 SECONDS

DEPTH BELOW M.L. (IN)	BENDING ALLOWABLE (KSI)	ACTUAL (KSI)	SHEAR ALLOWABLE (KSI)	ACTUAL (KSI)	COMBINED STRESS RATIO
.0	23.80	7.05	14.40	229.94	255.269
36.0	23.80	56.41	14.40	223.01	242.221
72.0	23.80	105.77	14.40	213.25	223.756
108.0	23.80	140.62	14.40	204.41	207.409
144.0	23.80	133.87	14.40	198.50	195.650
180.0	23.80	109.97	14.40	195.99	189.873
216.0	23.80	78.80	14.40	196.29	189.122
252.0	23.80	48.47	14.40	198.32	191.703
288.0	23.80	24.03	14.40	200.99	195.826
324.0	23.80	7.36	14.40	203.51	200.034
360.0	23.80	-2.01	14.40	205.42	203.588
396.0	23.80	-5.82	14.40	206.62	206.131
432.0	23.80	-6.14	14.40	207.20	207.297
468.0	23.80	-4.79	14.40	207.34	207.531
504.0	23.80	-3.01	14.40	207.25	207.273
540.0	23.80	-1.50	14.40	207.08	206.854
576.0	23.80	-.51	14.40	206.90	206.468
612.0	23.80	-.02	14.40	206.78	206.200
648.0	23.80	.11	14.40	206.72	206.077
684.0	23.80	.06	14.40	206.71	206.071
720.0	23.80	.00	14.40	206.73	206.109

PILE # 7 /SOIL LOADS AND DEFORMATIONS AT TIME =   .43897SECONDS

DEPTH BELOW M.L. (IN)	SOIL MODULUS (K/SQ IN)	SOIL RESISTANCE ALLOW. (K/IN)	ACTUAL (K/IN)	PILE 7 DEFL. DEFL (IN)	SLOPE (RAD)	PILE 7 M AND V MOMENT 0IN-K)	SHEAR (K)
TOP	.00	.0	.0	1.507	.005	0.	1984.5
.0	.00	.0	.0	.117	.001	752.	1973.4
36.0	21.60	233.0	1.9	.086	.001	6005.	1914.0
72.0	43.20	465.9	2.5	.057	.001	11259.	1830.2
108.0	64.80	698.9	2.2	.034	.001	14968.	1754.3
144.0	86.40	931.8	1.4	.016	.000	14249.	1703.6
180.0	108.00	1164.8	.6	.005	.000	11706.	1682.0
216.0	129.60	1397.7	-.1	-.001	.000	8387.	1684.6
252.0	151.20	1630.7	-.5	-.003	.000	5159.	1702.0
288.0	172.80	1863.6	-.7	-.004	-.000	2558.	1724.9
324.0	194.40	2096.6	-.6	-.003	-.000	783.	1746.5
360.0	216.00	2329.6	-.5	-.002	-.000	-214.	1763.0
396.0	237.60	2562.5	-.3	-.001	-.000	-619.	1773.3
432.0	259.20	2795.5	-.1	-.000	-.000	-654.	1778.2
468.0	280.80	3028.4	-.0	-.000	-.000	-510.	1779.5
504.0	302.40	3261.4	.0	.000	-.000	-321.	1778.7
540.0	324.00	3494.3	.0	.000	-.000	-160.	1777.2
576.0	345.60	3727.3	.0	.000	.000	-54.	1775.7
612.0	367.20	3960.3	.0	.000	.000	-2.	1774.6
648.0	388.80	4193.2	.0	.000	.000	12.	1774.1
684.0	410.40	4426.2	.0	.000	.000	6.	1774.1
720.0	432.00	4659.1	-.0	-.000	.000	0.	1774.2

DOLPHIN/SOIL LOADS AND DEFORMATIONS AT TIME =   .43897 SECONDS

DEPTH BELOW M.L. (IN)	BENDING ALLOWABLE (KSI)	ACTUAL (KSI)	SHEAR ALLOWABLE (KSI)	ACTUAL (KSI)	COMBINED STRESS RATIO
.0	23.80	7.07	14.40	229.94	255.269
36.0	23.80	56.43	14.40	223.01	242.218
72.0	23.80	105.80	14.40	213.25	223.750
108.0	23.80	140.65	14.40	204.40	207.400
144.0	23.80	133.90	14.40	198.50	195.639
180.0	23.80	110.00	14.40	195.99	189.861
216.0	23.80	78.81	14.40	196.28	189.110
252.0	23.80	48.48	14.40	198.31	191.692
288.0	23.80	24.04	14.40	200.99	195.816
324.0	23.80	7.36	14.40	203.50	200.024
360.0	23.80	-2.01	14.40	205.42	203.579
396.0	23.80	-5.82	14.40	206.62	206.123
432.0	23.80	-6.14	14.40	207.20	207.289
468.0	23.80	-4.79	14.40	207.34	207.523
504.0	23.80	-3.01	14.40	207.25	207.265
540.0	23.80	-1.50	14.40	207.07	206.846
576.0	23.80	-.51	14.40	206.90	206.460
612.0	23.80	-.02	14.40	206.77	206.192
648.0	23.80	.11	14.40	206.71	206.063
684.0	23.80	.06	14.40	206.71	206.069
720.0	23.80	.00	14.40	206.73	206.101

PILE # 8 /SOIL LOADS AND DEFORMATIONS AT TIME =   .43897SECONDS

DEPTH BELOW M.L. (IN)	SOIL MODULUS (K/SQ IN)	SOIL RESISTANCE ALLOW. (K/IN)	SOIL RESISTANCE ACTUAL (K/IN)	PILE 8 DEFL. DEFL (IN)	PILE 8 DEFL. SLOPE (RAD)	PILE 8 M AND V MOMENT (IN-K)	PILE 8 M AND V SHEAR (K)
TOP	.00	.0	.0	1.507	.005	0.	1984.5
.0	.00	.0	.0	.117	.001	753.	1973.4
36.0	21.60	233.0	1.9	.086	.001	6008.	1914.0
72.0	43.20	465.9	2.5	.057	.001	11262.	1830.2
108.0	64.80	698.9	2.2	.034	.001	14971.	1754.2
144.0	86.40	931.8	1.4	.016	.000	14252.	1703.5
180.0	108.00	1164.8	.6	.005	.000	11708.	1682.0
216.0	129.60	1397.7	-.1	-.001	.000	8389.	1684.5
252.0	151.20	1630.7	-.5	-.003	.000	5160.	1701.9
288.0	172.80	1863.6	-.7	-.004	-.000	2559.	1724.9
324.0	194.40	2096.6	-.6	-.003	-.000	783.	1746.5
360.0	216.00	2329.6	-.5	-.002	-.000	-214.	1762.9
396.0	237.60	2562.5	-.3	-.001	-.000	-619.	1773.2
432.0	259.20	2795.5	-.1	-.000	-.000	-654.	1778.2
468.0	280.80	3028.4	-.0	-.000	-.000	-510.	1779.4
504.0	302.40	3261.4	.0	.000	-.000	-321.	1778.7
540.0	324.00	3494.3	.0	.000	-.000	-160.	1777.1
576.0	345.60	3727.3	.0	.000	.000	-54.	1775.7
612.0	367.20	3960.3	.0	.000	.000	-2.	1774.6
648.0	388.80	4193.2	.0	.000	.000	12.	1774.0
684.0	410.40	4426.2	.0	.000	.000	6.	1774.0
720.0	432.00	4659.1	-.0	-.000	.000	0.	1774.2

DOLPHIN/SOIL LOADS AND DEFORMATIONS AT TIME =   .43897 SECONDS

DEPTH BELOW M.L. (IN)	BENDING ALLOWABLE (KSI)	BENDING ACTUAL (KSI)	SHEAR ALLOWABLE (KSI)	SHEAR ACTUAL (KSI)	COMBINED STRESS RATIO
.0	23.80	7.08	14.40	229.94	255.269
36.0	23.80	56.45	14.40	223.01	242.215
72.0	23.80	105.83	14.40	213.25	223.744
108.0	23.80	140.68	14.40	204.40	207.391
144.0	23.80	133.93	14.40	198.49	195.628
180.0	23.80	110.02	14.40	195.98	189.850
216.0	23.80	78.83	14.40	196.28	189.098
252.0	23.80	48.49	14.40	198.30	191.681
288.0	23.80	24.04	14.40	200.98	195.806
324.0	23.80	7.36	14.40	203.50	200.014
360.0	23.80	-2.01	14.40	205.41	203.570
396.0	23.80	-5.82	14.40	206.61	206.114
432.0	23.80	-6.14	14.40	207.19	207.281
468.0	23.80	-4.79	14.40	207.34	207.515
504.0	23.80	-3.01	14.40	207.25	207.257
540.0	23.80	-1.50	14.40	207.07	206.838
576.0	23.80	-.51	14.40	206.89	206.452
612.0	23.80	-.02	14.40	206.77	206.183
648.0	23.80	.11	14.40	206.71	206.060
684.0	23.80	.06	14.40	206.70	206.054
720.0	23.80	.00	14.40	206.73	206.092

PILE # 9 /SOIL LOADS AND DEFORMATIONS AT TIME = .43897 SECONDS

DEPTH BELOW M.L. (IN)	SOIL MODULUS (K/SQ IN)	SOIL RESISTANCE ALLOW. (K/IN)	ACTUAL (K/IN)	PILE 9 DEFL. DEFL (IN)	SLOPE (RAD)	PILE 9 M AND V MOMENT (IN-K)	SHEAR (K)
TOP .0	.00	.0	.0	-1.154	-.004	0.	1984.5
.0	.00	.0	.0	-.089	-.001	-210.	1993.0
36.0	21.60	233.0	-1.4	-.065	-.001	-4284.	2038.2
72.0	43.20	465.9	-1.9	-.044	-.001	-8358.	2102.1
108.0	64.80	698.9	-1.7	-.026	-.000	-11255.	2160.2
144.0	86.40	931.8	-1.1	-.013	-.000	-10771.	2199.2
180.0	108.00	1164.8	-.4	-.004	-.000	-8881.	2216.1
216.0	129.60	1397.7	.1	.001	-.000	-6365.	2214.5
252.0	151.20	1630.7	.4	.003	-.000	-3943.	2201.6
288.0	172.80	1863.6	.5	.003	-.000	-1968.	2184.3
324.0	194.40	2096.6	.5	.002	.000	-615.	2167.9
360.0	216.00	2329.6	.4	.002	.000	149.	2155.4
396.0	237.80	2562.5	.2	.001	.000	463.	2147.6
432.0	259.20	2795.5	.1	.000	.000	494.	2143.7
468.0	280.80	3028.4	.0	.000	.000	387.	2142.8
504.0	302.40	3261.4	-.0	-.000	.000	244.	2143.3
540.0	324.00	3494.3	-.0	-.000	.000	123.	2144.5
576.0	345.60	3727.3	-.0	-.000	-.000	42.	2145.6
612.0	367.20	3960.3	-.0	-.000	-.000	2.	2146.4
648.0	388.80	4193.2	-.0	-.000	-.000	-9.	2146.8
684.0	410.40	4426.2	-.0	-.000	-.000	-5.	2146.8
720.0	432.00	4659.1	.0	.000	-.000	0.	2146.7

DOLPHIN/SOIL LOADS AND DEFORMATIONS AT TIME = .43897 SECONDS

DEPTH BELOW M.L. (IN)	BENDING ALLOWABLE (KSI)	ACTUAL (KSI)	SHEAR ALLOWABLE (KSI)	ACTUAL (KSI)	COMBINED STRESS RATIO
.0	23.80	-1.97	14.40	232.22	260.145
36.0	23.80	-40.26	14.40	237.49	273.683
72.0	23.80	-78.54	14.40	244.93	292.611
108.0	23.80	-105.77	14.40	251.70	309.963
144.0	23.80	-101.21	14.40	256.25	320.909
180.0	23.80	-83.46	14.40	258.21	325.035
216.0	23.80	-60.00	14.40	258.03	323.607
252.0	23.80	-37.06	14.40	256.52	318.902
288.0	23.80	-18.49	14.40	254.51	313.155
324.0	23.80	-5.78	14.40	252.60	307.960
360.0	23.80	1.40	14.40	251.15	304.234
396.0	23.80	4.35	14.40	250.23	302.143
432.0	23.80	4.64	14.40	249.78	301.083
468.0	23.80	3.63	14.40	249.67	300.764
504.0	23.80	2.30	14.40	249.74	300.868
540.0	23.80	1.15	14.40	249.87	301.142
576.0	23.80	.40	14.40	250.00	301.425
612.0	23.80	.02	14.40	250.09	301.637
648.0	23.80	-.08	14.40	250.14	301.757
684.0	23.80	-.04	14.40	250.14	301.760
720.0	23.80	.00	14.40	250.13	301.721

PILE #10 /SOIL LOADS AND DEFORMATIONS AT TIME =   .43897 SECONDS

DEPTH BELOW M.L. (IN)	SOIL MODULUS (K/SQ IN)	SOIL RESISTANCE ALLOW. (K/IN)	ACTUAL (K/IN)	PILE 10 DEFL. DEFL (IN)	SLOPE (RAD)	PILE 10 M AND V MOMENT (IN-K)	SHEAR (K)
TOP	.00	.0	.0	1.327	.005	0.	1984.5
.0	.00	.0	.0	.102	.001	339.	1974.8
36.0	21.60	233.0	1.6	.075	.001	5009.	1922.8
72.0	43.20	465.9	2.2	.050	.001	9679.	1849.2
108.0	64.80	698.9	1.9	.030	.000	12995.	1782.4
144.0	86.40	931.8	1.3	.015	.000	12420.	1737.6
180.0	108.00	1164.8	.5	.005	.000	10232.	1718.4
216.0	129.60	1397.7	-.1	-.001	.000	7350.	1720.2
252.0	151.20	1630.7	-.5	-.003	.000	4535.	1735.2
288.0	172.80	1863.6	-.6	-.003	-.000	2260.	1755.2
324.0	194.40	2096.6	-.5	-.003	-.000	703.	1774.0
360.0	216.00	2329.6	-.4	-.002	-.000	-176.	1788.4
396.0	237.60	2562.5	-.2	-.001	-.000	-535.	1797.5
432.0	259.20	2795.5	-.1	-.000	-.000	-569.	1801.8
468.0	280.80	3028.4	-.0	-.000	-.000	-446.	1803.0
504.0	302.40	3261.4	.0	-.000	-.000	-281.	1802.3
540.0	324.00	3494.3	.0	.000	-.000	-141.	1801.0
576.0	345.60	3727.3	.0	.000	-.000	-48.	1799.7
612.0	367.20	3960.3	.0	.000	.000	-2.	1798.8
648.0	388.80	4193.2	.0	.000	.000	10.	1798.3
684.0	410.40	4426.2	.0	.000	.000	5.	1798.3
720.0	432.00	4659.1	-.0	-.000	.000	0.	1798.4

DOLPHIN/SOIL LOADS AND DEFORMATIONS AT TIME =   .43897 SECONDS

DEPTH BELOW M.L. (IN)	BENDING ALLOWABLE (KSI)	ACTUAL (KSI)	SHEAR ALLOWABLE (KSI)	ACTUAL (KSI)	COMBINED STRESS RATIO
.0	23.80	3.19	14.40	230.10	255.465
36.0	23.80	47.07	14.40	224.03	244.027
72.0	23.80	90.96	14.40	215.47	227.712
108.0	23.80	122.12	14.40	207.69	213.142
144.0	23.80	116.71	14.40	202.46	202.589
180.0	23.80	96.15	14.40	200.22	197.362
216.0	23.80	69.07	14.40	200.44	196.644
252.0	23.80	42.62	14.40	202.18	198.921
288.0	23.80	21.24	14.40	204.51	202.584
324.0	23.80	6.60	14.40	206.70	206.326
360.0	23.80	-1.65	14.40	208.38	209.478
396.0	23.80	-5.03	14.40	209.44	211.744
432.0	23.80	-5.35	14.40	209.95	212.790
468.0	23.80	-4.19	14.40	210.08	213.007
504.0	23.80	-2.64	14.40	210.00	212.785
540.0	23.80	-1.33	14.40	209.85	212.417
576.0	23.80	-.46	14.40	209.70	212.076
612.0	23.80	-.02	14.40	209.59	211.838
648.0	23.80	.10	14.40	209.53	211.728
684.0	23.80	.05	14.40	209.53	211.722
720.0	23.80	.00	14.40	209.55	211.755

TABLE 6

PAGE 52

OUTPUT OF DYNAMIC DATA AT THE POINT OF IMPACT

STEP NO	TIME (SEC)	DEFL (IN)	VELOCITY (IN/SEC)	ACCELERATION (IN/SEC/SEC)	FORCE (K.)	SPRING K (K/IN)
1	.0000	.00	6.48	.00	1.00	594.34
2	.1097	.68	5.71	-14.08	1021.40	1495.47
3	.2195	1.20	3.57	-24.81	1799.22	1495.47
4	.3292	1.44	.59	-29.62	2148.00	1495.47
5	.4390	1.33	-2.54	-27.36	1984.55	1495.47

OUTPUT OF KINETIC AND POTENTIAL ENERGY FOR SYSTEM

STEP NO	TIME (SEC.)	KINETIC ENERGY (IN-K)	POTENTIAL ENERGY (IN-K)	TOTAL PE + KE (IN--K)
1	.0000	1523.	0.	1523.
2	.1097	1182.	349.	1531.
3	.2195	463.	1083.	1546.
4	.3292	13.	1543.	1555.

```
        TOTAL KINETIC ENERGY OF SHIP =        1523. K-IN
   TOTAL POTENTIAL ENERGY OF DOLPHIN =        1543. K-IN
                            ERROR =          -1.30 PERCENT
```

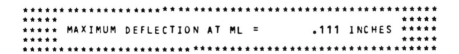

```
************************************************************
*****                                                 *****
***** MAXIMUM DEFLECTION AT ML =        .111 INCHES   *****
*****                                                 *****
************************************************************
```

FACTOR OF SAFETY =   139.609

EXAMPLE 4

General

The purpose of this example is to present the computer analysis of a bridge cell protective system. The bridge, as shown in Figure 19, is located north of St. Louis, Missouri; across the Illinois River near the city of Pearl, Illinois.

As described previously the method of analysis consists of a computer oriented dynamic response program, which can analyze various fendering schemes on single dolphins, clusters of cells.

The basic theory involves matrix formulation of interactive pilings, attached by cross members, which support energy absorbing fenders. Soil characteristics can be designated along the length of the embedded pilings. Isolated dolphins are similarly modeled, and can include wood clusters with cabled wrappings, rigid pile clusters or cells.

In general, the input requirements are the ship tonnage, speed, direction, rigidity of supporting elements, cantilever length, energy characteristics of fenders, soil parameters and general geometry of system. The output information includes velocity and acceleration of vessel at any time interval and the corresponding forces and deformations throughout the system.

BRIDGE DESCRIPTION

The replaced bridge consists of a 367 ft., vertical lift steel truss, which will provide a 69' vertical clearance to the shipping traffic. The previous bridge, which was a swing structure had a 149' clear channel. This new structure will provide a 315 ft clear channel, as shown in Figure 20.

PROTECTIVE CELL

In order to protect the new bridge piers, four steel cells, will be positioned near the bridge as shown in Figure 2. These cells consist of steel sheet piling (58 sheets/cell) with a 24'-7 3/8" diameter. The units are 61.0 ft. long and are filled with 2500 psi lean concrete, as shown in Figure 21. The cell will have a cantilever length of 36 ft, if the existing river bed is maintained and 41 ft. for the dredged river. The vessel, which will impact the cell, was assumed as follows and dictated by the USCG;

Wt. (tons)	Velocity (knots)
10,000	0.40
1,000	1.40

COMPUTER RESULTS

a) Systems Analyzed

Two basic systems were analyzed, one which consists of only the infill concrete, and the other the steel shell. The cantilever length was changed for the concrete unit in order to examine the effect. The basic model is shown in Figure 22, and shows the cantilever length of the cell and the embedment length. Although, the concrete will not be acting independent of the steel, the effect of its stiffness is generally neglected in evaluating the strength of the steel shell. It was for this reason that the region of the concrete mass was examined.

b) Case Studies

As shown in Table 7, five case solutions were conducted. Cases 1 through 3, consisted of only the 27' diameter concrete mass and cases 4 and 5 consisted of the steel shell. The variations within each type was due to vessel weight and velocity and cantilever lengths.

c) Computer Output

Using the various parameters required for each of the five cases, a computer solution was obtained. The resulting output, which consists of a description of the input parameters and the final dynamic data, is given in Table 9.

Examination of these data shows that the steel cell will deform a maximum of 3.83 inches and a maximum stress of 8.2 ksi. A summary of the data for all five cases is shown in Table 8.

193

V.   RECOMMENDATIONS

The computer response of the 1" thick-28' round steel cell indicates that it will perform adequately when impacted by a ship of 10,000. T (0.4 knots) or 1000.T(1.4 knots). The maximum deformation of the cell is 3.83 inches and develops a maximum steel stress of 8.2 ksi.

TABLE 7

ANALYTICAL STUDIES

Case No.	Description	Cantilever Length (ft)	Ship Details	
1	Concrete Mass	36	$10,000^T$	0.4 knots
2	Concrete Mass	36	$1,000^T$	1.4 knots
3	Concrete Mass	41	$1,000^T$	1.4 knots
4	Steel Shell – 1" thick	36	$1,000^T$	1.4 knots
5	Steel Shell – 1" thick	36	$10,000^T$	0.4 knots

TABLE 8

SUMMARY OF RESULTS

CASE	CONDITION	DEFL (in)	FORCE (kips)	MOMENT (k-in)	SECTION (in^3)	STRESS (ksi)	STRESS ULT (ksi)	STATUS
1	conc.	3.38	1115.0	522335.	2.56×10^6	.204	.400	OK
2	conc.	3.74	1234.0	578162.	2.56×10^6	.226	.400	OK
3	conc.	5.98	772.	398077.	2.56×10^6	.155	.400	OK
4	cell	3.83	1206.	564822.	6.9×10^4	8.2	36.	OK
5	cell	3.44	1090.	510322.	6.9×10^4	7.4	36.	OK

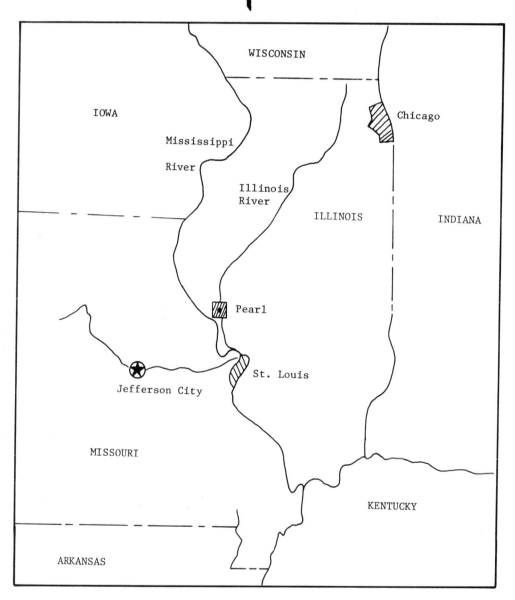

BRIDGE LOCATION

FIGURE 19

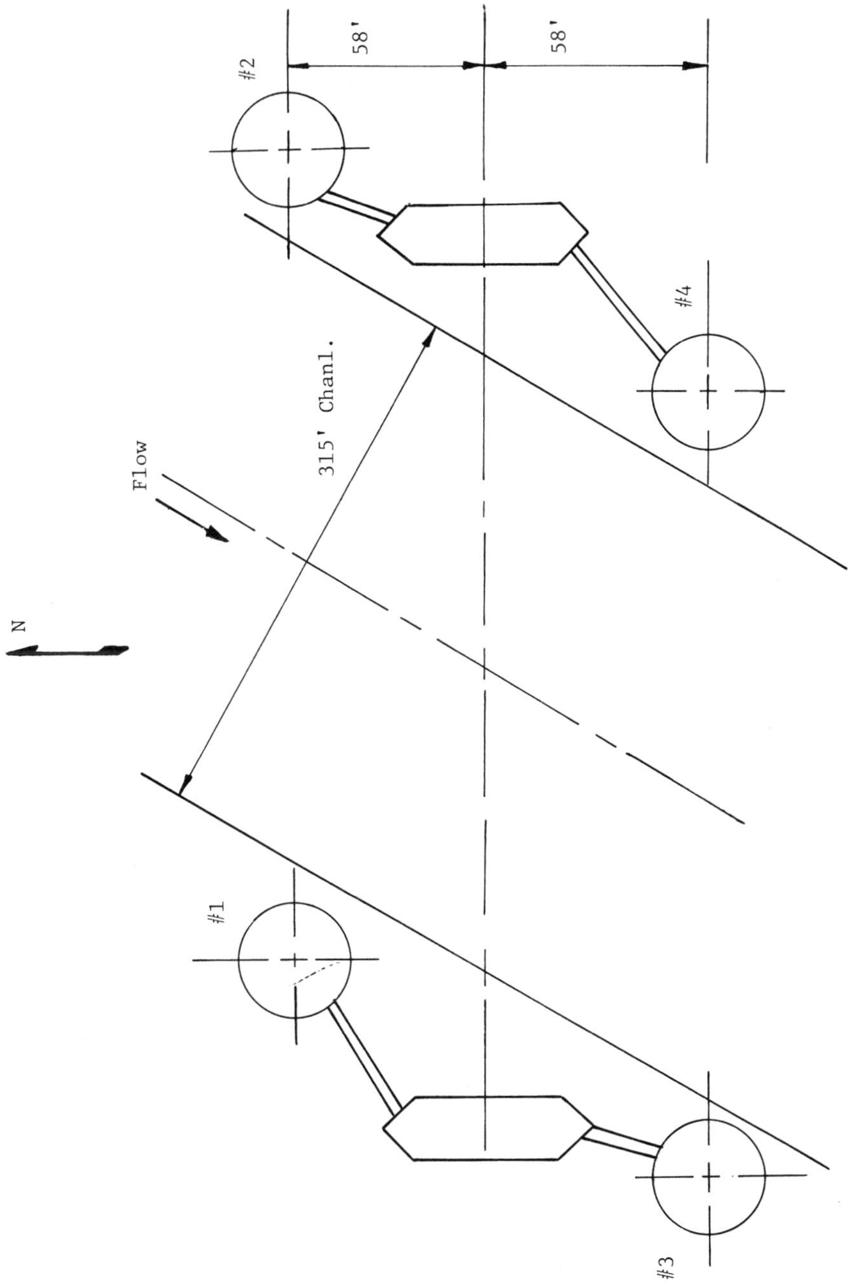

FIGURE 20

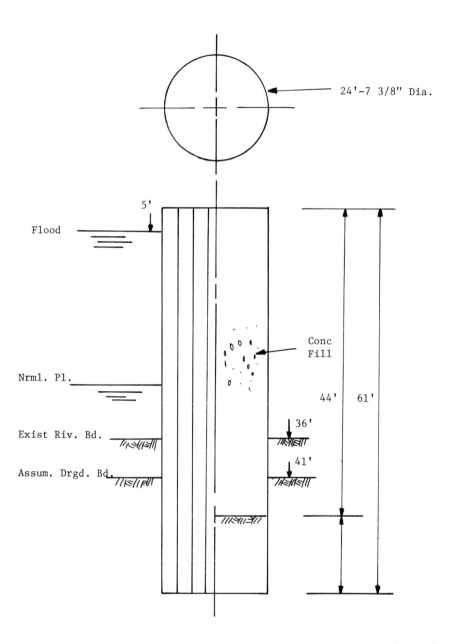

24'-7 3/8" Dia.

Flood

5'

Conc
Fill

Nrml. Pl.

44'   61'

Exist Riv. Bd.   36'

Assum. Drgd. Bd.   41'

FIGURE 21

198

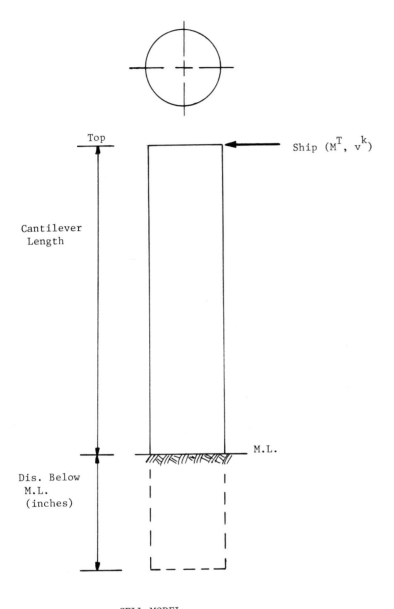

CELL MODEL

FIGURE 22

BASIC PROGRAM PARAMETERS

WEIGHT OF SHIP	=	10000.00 TONS (22400.00 KIPS)
VELOCITY OF SHIP	=	24.0 KNOTS (8.10 IPS)
DIAMETER OF DOLPHIN	=	.750 INCHES
WALL THICKNESS OF DOLPHIN	=	25.000 FEET
LENGTH OF DOLPHIN BELOW MUD	=	36.000 FEET
LENGTH OF DOLPHIN ABOVE SHIP	=	372863.16 IN**4
MOMENT OF INERTIA OF DOLPHIN	=	45.00 DEGREES
ANGLE OF PASSIVE FAILURE	=	.500 FEET
DISSIPATION CORRECTION FACTOR	=	.0030 IN/IN
DEPTH CORRECTION FACTOR	=	.0449 INCHES
ALLOWABLE ELASTIC STRAIN	=	.0029 SEC
ALLOWABLE DEFLECTION	=	
STEPSIZE	=	

DYNAMIC RESULTS FOR LINEAR SPRINGING

MAX. DEFLECTION AT POINT OF IMPACT	=	3.419 INCHES
MAXIMUM SHIP ACCELERATION	=	19.197 IN/SEC**2
MAXIMUM SHIP FORCE	=	11782 IKIPS
STOPPING TIME	=	.1620 SECONDS
INITIAL SPRING CONSTANT	=	0.78 IK/IN
LINEAR LAMBDA FACTOR	=	2.696 1/SEC

TABLE 9

DOLPHIN/SOIL LOADS AND DEFORMATIONS AT TIME = .72919 SECONDS

DEPTH BELOW TOP (IN)	SOIL MODULUS (K/SQ IN)	SOIL ALLOW (K/IN)	RESISTANCE ACTUAL (K/IN)	DOLPHIN DEFL (IN)	DOLPHIN DEFL. SLOPE (RAD)	DOLPHIN MOMENT (IN-K)	DOLPHIN M AND V SHEAR (K)

OUTPUT OF DYNAMIC DATA AT THE POINT OF IMPACT

STEP NO	TIME (SEC)	DEFL (IN)	VELOCITY (IN/SEC)	ACCELERATION (IN/SEC/SEC)	FORCE (K)	SPRING K (K/IN)

OUTPUT OF KINETIC AND POTENTIAL ENERGY FOR SYSTEM

STEP NO	TIME (SEC.)	KINETIC ENERGY (IN-K)	POTENTIAL ENERGY (IN-K)	TOTAL PE + KE (IN-K)
1	.0003	1904.	0.	1904.
2	.0005	1905.	47.	1945.
3	.0009	1822.	183.	1906.
4	.0011	1547.	359.	1907.
5	.0012	1254.	652.	1907.
6	.0024	960.	947.	1907.
7	.0040	694.	1213.	1908.
8	.0369	404.	1503.	1908.
9	.5069	183.	1725.	1908.
10	.5069	47.	1861.	1908.
11	.6629	0.	1908.	1908.

TOTAL TOTAL KINETIC ENERGY OF SHIP = 1004. K-IN
TOTAL POTENTIAL ENERGY OF DOLPHIN = 1608. K-IN
ERROR = -.22 PERCENT

****
**   **
**   MAXIMUM DEFLECTION AT ML = 1.095 INCHES **
**   **
****

FACTOR OF SAFETY = 24.111

BASIC PROGRAM PARAMETERS
=-=-=-=-=-=-=-=-=-=-=-=-=

WEIGHT OF SHIP   = 1000.00 TONS   (2240.00 KIPS)
VELOCITY OF SHIP   = 14.00 KNOTS   (28.36 IPS)
DIAMETER OF DOLPHIN   = 24.000 FEET INCHES
LENGTH OF DOLPHIN   = .000 FEET
LENGTH OF DOLPHIN   = 60.000 FEET

MOMENT OF INERTIA OF DOLPHIN = 378865.16 KIPS/IN**4
ANGLE OF PASSIVE FAILURE   = 45.000 DEGREES
DISSIPATION CORRECTION FACTOR = .00 FEET
DEPTH OF STRAIN   = .030000 IN/IN

ALLOWABLE ELASTIC DEFLECTION = .020485 INCHES
STEP SIZE   = .020000 SEC

DYNAMIC RESULTS FOR LINEAR SPRINGING
=-=-=-=-=-=-=-=-=-=-=-=-=-=-=-=-=-=-=

MAX. DEFLECTION AT POINT OF IMPACT = 3.784 INCHES
MAXIMUM SHIP ACCELERATION   = 212.4207 IN/SEC**2
MAXIMUM SHIPPING FORCE   = 1230.07 IN KIPS.
INITIAL SPRING STOPPING TIME   = .025.6 SECONDS
INITIAL SPRING CONSTANT   = 225.783 KSEK/IN
LINEAR LAMBDA FACTOR   = 7.4933 1/SEC

DOLPHIN/SOIL LOADS AND DEFORMATIONS AT TIME = .27059 SECONDS

DEPTH BELOW BML (IN)	SOIL MODULUS (K/SQ IN)	SOIL RESISTANCE		DOLPHIN DEFL.		DOLPHIN M AND V	
		ALLOW (K/IN)	ACTUAL (K/IN)	DEFL (IN)	SLOPE (RAD)	MOMENT (IN-K)	SHEAR (K)

OUTPUT OF DYNAMIC DATA AT THE POINT OF IMPACT

STEP NO	TIME (SEC)	DEFL (IN)	VELOCITY (IN/SEC)	ACCELERATION (IN/SEC/SEC)	FORCE (K)	SPRING K (K/IN)

OUTPUT OF KINETIC AND POTENTIAL ENERGY FOR SYSTEM

STEP NO	TIME (SEC.)	KINETIC ENERGY (IN-K)	POTENTIAL ENERGY (IN-K)	TOTAL PE + KE (IN-K)

TOTAL TOTAL KINETIC ENERGY OF SHIP = 2332. K-IN
TOTAL POTENTIAL ENERGY OF DOLPHIN = 2338. K-IN
                            ERROR = -.22 PERCENT

****
****
**   MAXIMUM DEFLECTION AT ML = 1.212 INCHES **
****
****

FACTOR OF SAFETY = 24.111

BASIC PROGRAM PARAMETERS

```
WEIGHT       OF SHIP          =   1000.00  TONS    (    2240.00 KIPS)
VELOCITY     OF SHIP          =     14.0   KNOTS   (      28.36 IPS)
DIAMETER     OF DOLPHIN       =     24.000 FEET INCHES
THICKNESS    OF DOLPHIN       =     24.000 FEET
LENGTH       OF DOLPHIN ABOVE =     47.0.. FEET
LENGTH       OF DOLPHIN BELOW =     24.00  DEGREES

MOMENT OF INERTIA OF DOLPHIN  = 378663416.00  IN**4
ANGLE OF PASSIVE FAILURE      =     45.00  DEGREES
     DISSIPATION CORR FACTOR  =       .50000  IN/IN
ALLOWABLE ELASTIC STRAIN      =       .03000  IN/IN
     DEFLECTION STEP SIZE     =       .035   INCHES
```

DYNAMIC RESULTS FOR LINEAR SPRINGING

```
MAX. DEFLECTION AT POINT OF IMPACT =   132.049  INCHES
     MAXIMUM SHIP ACCELERATION     =     6.912  IN/SEC**2
     MAXIMUM SHIP FORCE            =     .3351  KIPS
          STOPPING TIME            =     .3351  SECONDS
     INITIAL SPRING CONSTANT       =   147.48   KIP/IN
     LINEAR LAMBDA FACTOR          =     4.5674  1/SEC
```

DOLPHIN/SOIL LOADS AND DEFORMATIONS AT TIME = .36662 SECONDS

DEPTH BELOW (IN.) TOP	SOIL MODULUS (K/SQ IN)	SOIL RESISTANCE ALLOW (K/IN)	SOIL RESISTANCE ACTUAL (K/IN)	DOLPHIN DEFL (IN)	DOLPHIN DEFL SLOPE (RAD)	DOLPHIN MOMENT (IN-K)	DOLPHIN M AND V SHEAR (K)

209

OUTPUT OF DYNAMIC DATA AT THE POINT OF IMPACT

STEP NO	TIME (SEC)	DEFL (IN)	VELOCITY (IN/SEC)	ACCELERATION (IN/SEC/SEC)	FORCE (K)	SPRING K (K/IN)
1	0.000	0.0	6.7	20.5	0.0	18.472
2	0.003	0.7	6.0	10.2	1.4	18.472
3	0.006	2.5	5.7	2.7	3.3	19.472
4	0.009	3.5	5.1	5.4	4.3	19.472
5	0.013	4.4	4.2	9.6	5.2	18.472
6	0.016	5.2	3.0	12.8	6.0	19.472
7	0.020	5.9	2.0	−15.1	6.7	18.472
8	0.023	6.3	0.9	−17.0	7.2	19.472
9	0.026	6.8	−0.2	−18.5	7.7	19.472
10	0.030	6.8	−1.4	−19.1	7.7	18.472
11	0.033	6.4	−2.5	−13.3	7.2	19.472
12	0.036	5.6	−	−4	6.0	18.472

OUTPUT OF KINETIC AND POTENTIAL ENERGY FOR SYSTEM

STEP NO	TIME (SEC.)	KINETIC ENERGY (IN-K)	POTENTIAL ENERGY (IN-K)	TOTAL PE + KE (IN-K)

TOTAL TOTAL KINETIC ENERGY OF SHIP = 2732. K-IN
TOTAL POTENTIAL ENERGY OF DOLPHIN = 2738. K-IN
ERROR = -.27 PERCENT

MAXIMUM DEFLECTION AT ML = 1.509 INCHES

FACTOR OF SAFETY = 16.744

211

BASIC PROGRAM PARAMETERS

```
                WEIGHT OF SHIP  = 1000.00 TONS  (  2240.00 KIPS)
              VELOCITY OF SHIP  =   14.00 KNOTS (    28.36 IPS)
           DIAMETER OF DOLPHIN  =   24.000 FEET
     WALL THICKNESS OF DOLPHIN  =    1.0000 INCHES
     LENGTH OF DOLPHIN BELOW MUD =  36.000 FEET
     LENGTH OF DOLPHIN ABOVE MUD =  36.000 FEET
     MOMENT OF INERTIA OF DOLPHIN = 192257.34 KIN**4
        ANGLE OF PASSIVE FAILURE =  45.00 DEGREES
              DEPTH CORRECTION    =         FEET
           DISSIPATION FACTOR     =  .5000 IN/IN
     ALLOWABLE ELASTIC STRAIN     =  .03000 IN/IN
     ALLOWABLE DEFLECTION         =  .02485 INCHES
              STEPSIZE            =  .02145 SEC
```

DYNAMIC RESULTS FOR LINEAR SPRINGING

```
MAX. DEFLECTION AT POINT OF IMPACT  = 3.871 INCHES
         MAXIMUM SHIP ACCELERATION  = 207.03 IN/SEC**2
              MAXIMUM SHIP FORCE     = .605 IN/SEC
                SHIP STOPPING FORCE  = 12014.03 KIPS
                   STOPPING TIME     = .145 SECONDS
           INITIAL SPRING CONSTANT   = 211.23 KX/IN
             LINEAR LAMBDA FACTOR    = 7.246 1/SEC
```

212

DOLPHIN/SOIL LOADS AND DEFORMATIONS AT TIME = .23590 SECONDS

DEPTH BELOW GM.L. (IN)	SOIL MODULUS (K/SQ IN)	SOIL RESISTANCE ALLOW. (K/IN)	ACTUAL (K/IN)	DOLPHIN DEFL. DEFL (IN)	SLOPE (RAD)	DOLPHIN M AND V MOMENT (IN-K)	SHEAR (K)

OUTPUT OF DYNAMIC DATA AT THE POINT OF IMPACT

STEP NO	TIME (SEC)	DEFL (IN)	VELOCITY (IN/SEC)	ACCELERATION (IN/SEC/SEC)	FORCE (K)	SPRING K (K/IN)
1						
2						
3						
4						
5						
6						
7						
8						
9						
10						
11						
12						

214

OUTPUT OF KINETIC AND POTENTIAL ENERGY FOR SYSTEM

STEP NO	TIME (SEC.)	KINETIC ENERGY (IN-K)	POTENTIAL ENERGY (IN-K)	TOTAL PE + KE (IN-K)

TOTAL TOTAL KINETIC ENERGY OF SHIP = 2332. K-IN
TOTAL POTENTIAL ENERGY OF DOLPHIN = 2338. K-IN
ERROR = -.22 PERCENT

***** MAXIMUM DEFLECTION AT ML = 1.196 INCHES *****

FACTOR OF SAFETY = 24.111

215

```
BASIC PROGRAM PARAMETERS

        WEIGHT OF SHIP          =  10300.00 TONS  ( 22400.00 KIPS)
        VELOCITY OF SHIP        =     24.0 KNOTS  (     8.10 IPS)
        DIAMETER OF DOLPHIN     =     24.000 FEET
        WALL THICKNESS OF DOLPHIN =    1.5.00 INCHES
        LENGTH OF DOLPHIN BELOW MLP =  26.00 FEET
        LENGTH OF DOLPHIN ABOVE MLP =  36.00 FEET

        MOMENT OF INERTIA OF DOLPHIN =  1.225E+02 KIPS/IN**4
        ANGLE OF PASSIVE FAILURE     =    45.00 DEGREES
        DISSIPATION CORRECTION FACTOR =      .50 FEET
        ALLOWABLE STRAIN             =   .0200 IN/IN
        ELASTIC DEFLECTION            =   .45 INCHES
        STEPSIZE                     =  .06732 SEC


DYNAMIC RESULTS FOR LINEAR SPRINGING

MAX. DEFLECTION AT POINT OF IMPACT   =  18.498 INCHES
     MAXIMUM SHIP ACCELERATION        =  3.650 IN/SEC**2
     MAXIMUM SHIP FORCE               = 18.875 INKIPS
     INITIAL SHIP STOPPING TIME       =   .0752 SECONDS
     SPRING CONSTANT                  = 18.082 KIPS/IN
     LINEAR LAMBDA FACTOR             =  2.3162 1/SEC
```

DOLPHIN/SOIL LOADS AND DEFORMATIONS AT TIME = .74598 SECONDS

DEPTH BELOW BM-L (IN)	SOIL MODULUS (K/SQ IN)	SOIL RESISTANCE ALLOW (K/IN)	ACTUAL (K/IN)	DOLPHIN DEFL. DEFL (IN)	SLOPE (RAD)	DOLPHIN M AND V MOMENT (IN-K)	SHEAR (K)

OUTPUT OF DYNAMIC DATA AT THE POINT OF IMPACT

STEP NO	TIME (SEC)	DEFL (IN)	VELOCITY (IN/SEC)	ACCELERATION (IN/SEC/SEC)	FORCE (K)	SPRING K (K/IN)
1						
2						
3						
4						
5						
6						
7						
8						
9						
10						
11						
12						

OUTPUT OF KINETIC AND POTENTIAL ENERGY FOR SYSTEM

STEP NO	TIME (SEC.)	KINETIC ENERGY (IN-K)	POTENTIAL ENERGY (IN-K)	TOTAL PE + KE (IN-K)

TOTAL KINETIC ENERGY OF SHIP = 1904. K-IN
TOTAL POTENTIAL ENERGY OF DOLPHIN = 1905. K-IN
ERROR = -.22 PERCENT

MAXIMUM DEFLECTION AT ML = 1.033 INCHES

FACTOR OF SAFETY = 29.633

219

CHAPTER VII

CONCLUSIONS

The results of this study have led to several major conclusions:

1. The majority of fendering systems used in this country consist of pile or rubber fenders or a combination of the two. These systems were generally designed by experience rather than by detailed energy considerations resulting in both over and under design.

2. The variety of fendering systems available allows design control over a wide range of stiffness and energy criterion. Combining several systems in one installation may yield a more appropriate design than a homogeneous system.

3. Wood is the most commonly used fendering material due to its high fiber strength and hardness, its resilence, its relative abundance and low cost. Whenever wood is used in the marine environment, it must be protected by some treatment preventing borer, termite, and ant attack as well as reducing the abrasive effects of sand, silt, and ice.

4. Steel sections used in fendering systems must be heavily protected against corrosion and abrasion. Shotcrete and gunite have been used in the past, but the development of plastic, acrylic, or resin coatings may be preferrable. Steel sections are used where long lengths, low displacement, and high strength are required.

5. Concrete structures tend to be massive and too stiff for many vessels. Rubber or timber rubbing strips are necessary

on exposed concrete faces. The concrete mix should have a cement content between 6-1/2 sacks/cu. yd. and 7-1/2 sacks/cu. yd. The aggregates should be graded for maximum density and non-reactive. The water content should be the lowest possible to provide a workable, plastic mix, but not to exceed six gallons/sack of cement. Type V or Type II cements are preferrable and three percent to six percent entrained air is recommended to increase abrasion resistance and workability. Adequate vibration is necessary to ensure concrete strength. Minimum depth of covering over reinforcement should be three inches.

6. Complete detailed design of a fender system and its support structure requires the use of a rigid mass and spring model. This process would be extremely difficult without the use of computers. This system is currently being developed.

7. In lieu of a computer package, hand calculations may be based on the following design parameters: approach angle and velocity, ship displacement, draft and beam, general configuration of the support system (open, partially open, closed), an assumed stiffness between the fender and support structure, impact point, and the stiffness and energy absorption characteristics of the fender. These calculations do not account for the distribution among piles of horizontal loads on a piling system. Nor do they consider the effects of soil deformation and the resulting shift in the fixed point of the pile system. They represent an improvement over past design methods and are generally adequate.

8. The factor of safety concept should not be applied to fender design as some arbitrary multiplicative constant. The system should be designed for a range of conditions with selection based on the most serious conditions. The non-linear nature of most of the stiffness curves will generally increase the ultimate energy of the system even further. Conditions which exceed the design parameters may result in failure and should be considered accidental occurrences, especially impact velocity.

9. Future research should deal with new fendering systems, corrosion protection of steel and concrete, development of computer simulation, and design velocities.

10. The above mentioned future research will also include a study of the rotational effect of the vessel after vessel impact. This would also include the shock absorbing effect of the hull movement.

CHAPTER VIII

RECOMMENDATIONS FOR FUTURE RESEARCH

It is obvious from the completed study that present-day bridge protection systems and devices are inadequate. In other words, they are either under or over-designed. This is attributed to the fact that tankers, containerships, and barge tows have increased in substantial size without a proportional change in design criteria or innovative ideas in the bridge protection area.

Future research should center on the design, analysis, and laboratory modeling of new and innovative ideas in fendering, protective cells and shear fences. This type of research should be incorporated into a Phase II study.

With today's inadequate bridge protective systems and devices it would be appropriate to select a specific bridge which has received considerable publicity because of its history of collisions, damages, and delays to navigation and conduct a model test in a laboratory. A typical example would be the Southern Pacific Transportation Company bridge across the Atchafalaya River at Berwick Bay, Morgan City, Louisiana. If an innovative bridge protective system or device could be developed through a model study for such a troublesome bridge as Berwick Bay, then the results of this study can be adapted to similar projects which are less hazardous.

APPENDIX I

GENERAL

The following pages contain tables and figures representing the stiffness and energy characteristics of particular fenders at given deflection levels. The energy tables are read by finding the fender name or size designation in the left hand column. The deflection levels are shown across the top of the tables as two digit numbers representing the percent of the maximum deflection (Δ_{max}). At the intersection of the row and column is the energy value. The units of the energy are found in the right hand margin. The maximum deflection (Δ_{max}) is expressed in inches unless otherwise indicated. The curve coefficient tables are read in a similar manner. These tables represent the coefficient values of a fitting function of some form which is given with each table. The deflection limits, in inches, for which the function applies is found under "upper" or "lower boundary" as appropriate. If more than one function is used for fitting, they are indicated. Finally, stiffness curves are presented for all fenders and are used to verify calculations or computed reaction forces or stiffnesses as appropriate.

APPENDIX I-A

BAKKER RUBBERFABRIEK B.V.

Table I-A-1: BAKKER ENERGY TABLE

Goliath	10	20	30	40	50	60	70	80	90	100	Δ_{max}	
800x 400					174	261	434	650	865	1216	15.7	E = K-in.
1000x 500					350	521	780	1130	1433	2000	20.5	
1200x 600				350	650	870	1303	1738	2170	2610	25.6	
1400x 800				300	475	695	950	1390	2000	3475	44.1	
1500x 800				300	475	695	1000	1564	2260	3475	43.3	
1600x 800				350	520	782	1433	2215	3300	4344	44.5	
1750x1000				1216	1650	2260	3130	3910	5213	6342	41.3	
1850x1000				1480	1910	2780	3475	4604	5950	8690	41.7	
2050x1050				1650	2085	2867	3736	4865	6973	8690	39.0	

Wing Type												
170x 100				4.0	4.8	6.6	9.3	13	26	50	2.26	E = K-in/
228x 139				3.1	5.0	6.1	8.5	13	26	66	4.04	ft. of length

Rubber Buffer Cyl-axial				7.9	12.	20	29	37	50	66	5.22	
21x10.5					213	230	390	565	652	825	13.8	E - K-in.
18x 9					209	221	326	434	539	652	13.8	
15x 7.5					110	170	217	278	374	440	12.2	
12x 6					55	104	110	190	217	228	10.2	

Shear												
200x 150					11	16	22	30	39	43	8.26	K-in kips
500x 250					22	35	48	61	74	96	8.86	Δ-horizon-
300x 150					30	42	55	67	87	109	8.66	tal dis-
400x 150					43	56	78	96	120	152	7.48	placement,
550x 200					56	78	109	140	161	220	7.28	inches

D-Fender												
4x 3.75					3.3	6.5	10	26	53	87	3.0	K-in.
8x 6					6.6	12	26	53	106	152	3.94	kips/foot
10x 8					13	27	37	65	119	205	5.41	
12x10					16	40	60	115	180	265	6.10	

Table I-A-2: BAKKER CURVE COEFFICIENTS

A   Goliath $K = a + b\Delta$ or $= a' + b'\Delta + c'\Delta^2$ K = Kips per inch

	a	b	boundary lower	upper	a'	b'	c'	boundary lower	upper
800x 400	6.20	+.10	8.0	10.0	24.54	-3.30	.156	10.0	16.75
1000x 500	2.932	+.071	8.0	14.0	36.58	-4.20	.152	14.0	21.20
1200x 600	7.0	0	8.0	17.0	34.89	-3.27	.096	17.0	25.1
1400x 800	---	---	---	---	2.248	- .074	.003	8.0	39.4
1500x 800	---	---	---	---	3.188	- .084	.003	8.0	39.4
1600x 800	3.50	0	8.0	18.0	6.107	- .270	.007	18.0	47.2
1750x1000	7.35	0	8.0	29.0	24.08	-1.17	.020	29.0	39.4
1850x1000	8.40	0	8.0	29.0	27.17	-1.29	.022	29.00	39.4
2050x1050	10.50	0	8.0	29.0	41.08	-2.13	.037	29.0	39.4

B   Wing Type $K = a + b\Delta$ or $= a' + b'\Delta + c'\Delta^2$ K = Kips per inch per foot

	a	b	boundary lower	upper	a'	b'	c'	boundary lower	upper
170x 110	---	---	---	---	11.73	-19.57	10.72	.78	2.00
228x 139	1.051	+.349	.75	3.15	69.41	-40.62	6.12	3.15	4.2

C   Rubber Buffer $K = a + b\Delta$ or $= a' + b'\Delta + c'\Delta$ K = Kips per inch per foot

a	b	boundary lower	upper	a'	b'	c'	boundary lower	upper
4.0	0	.75	3.15	5.25	- .820	.134	3.15	5.50

D   Bakker Cylinder Axially Loaded   K = Constant   K = Kips per inch

	k	lower	upper
21x10.5	8.6	2.0	11.8
18x 9	7.3	2.0	11.8
15x 7.5	6.0	2.0	11.8
12x 6	5.0	2.0	11.8

E   Shear Fender $K = a + b\Delta$

K = Kips per inch

	a	b	boundary lower	boundary upper
200x150	1.302	-.011	2.0	10.0
500x250	2.214	-.007	2.0	10.0
300x150	3.0	0	2.0	10.0
400x150	5.184	-.017	2.0	8.0
550x200	7.426	-.038	2.0	6.0

F   D-Fenders

$$K = a + b\Delta \text{ or}$$
$$= a' + b'\Delta + c'\Delta^2$$

K = Kips per inch per foot

	a	b	boundary lower	boundary upper	a'	b'	c'	boundary lower	boundary upper
4x3.75	---	---	---	---	34.72	-49.55	18.60	1.0	2.50
8x6	---	---	---	---	11.85	-13.31	5.39	1.0	4.0
10x8	5.2	0	1.0	2.0	16.20	-10.43	2.47	2.0	5.0
12x10	6.8	0	1.0	2.5	28.21	-14.92	2.54	2.5	6.0

Bakker Goliath Fenders

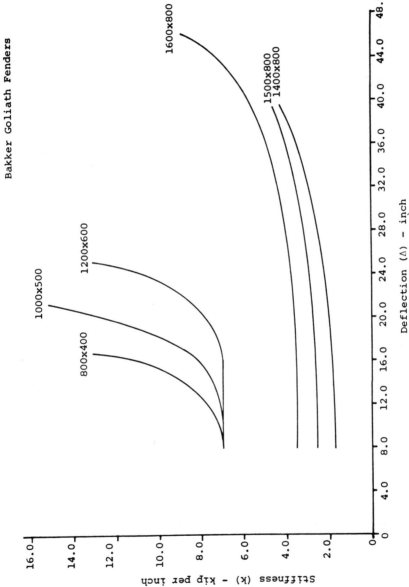

FIG. I-A-1: Stiffness vs. Deflection for Bakker Goliath Fenders up to 1600 x 800

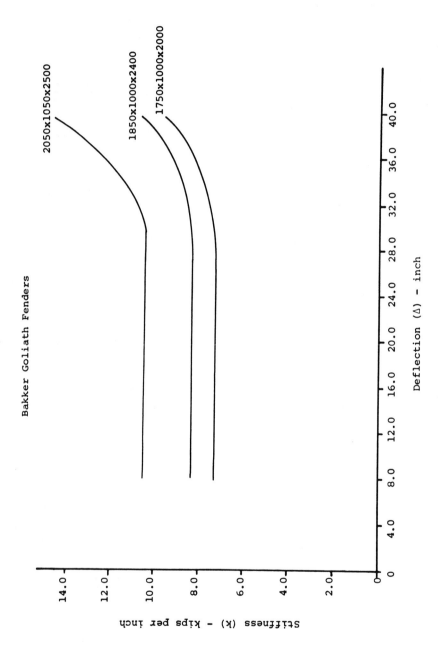

FIG. I-A-2: Stiffness vs. Deflection for Bakker Goliath up to 2050 x 1050 x 2500

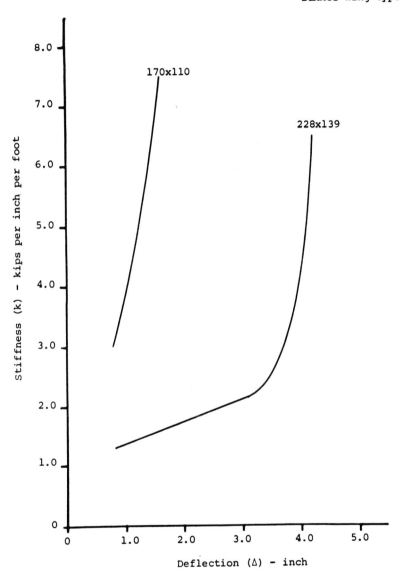

FIG. I-A-3: Stiffness vs. Deflection for Bakker Wing-type Fender

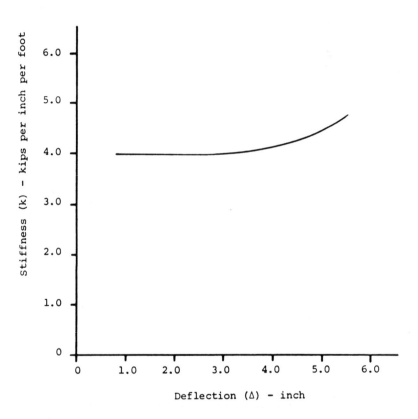

FIG. I-A-4: Stiffness vs. Deflection for Bakker Rubber Buffer Fender

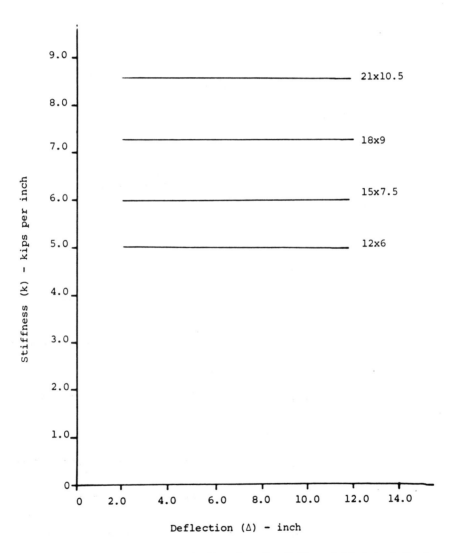

FIG. I-A-5: Stiffness vs. Deflection for Bakker Cylindrical Fender
         Axially Loaded

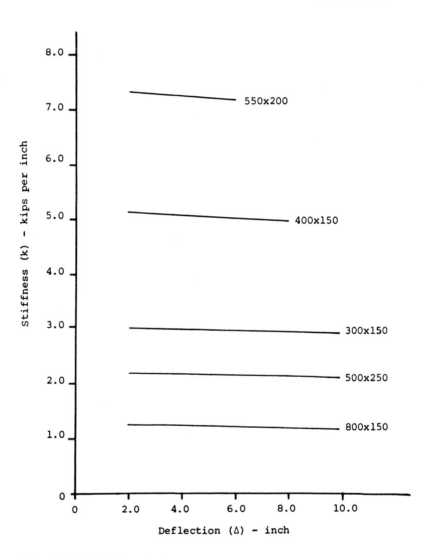

FIG. I-A-6: Stiffness vs. Deflection for Bakker Shear Fender

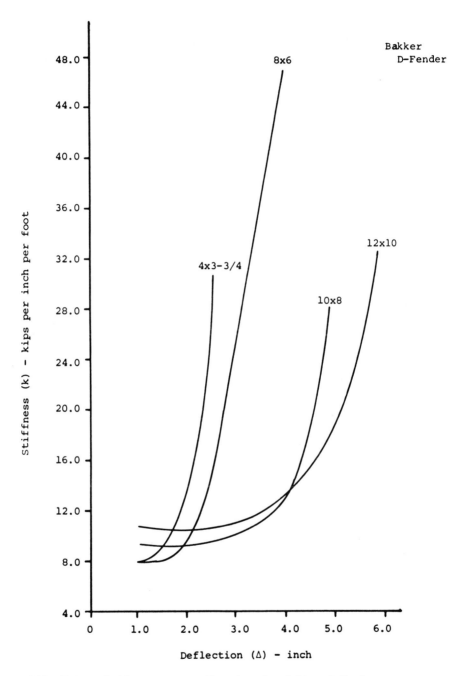

FIG. I-A-7: Stiffness vs. Deflection for Bakker D Fender

APPENDIX I-B

DUNLOP GRG DIVISION

Table I-B-1: DUNLOP ENERGY TABLE

	50	55	60	$\Delta_{\mathbf{max}}$
39" Dia.	54	78	113	E = Kip-in/per meter effective length
60"	117	182	325	
72"	165	243	375	
90"	278	400	591	
108"	382	570	817	
126"	540	773	1130	
174"	1086	1615	2354	

237

Table I-B-2: DUNLOP CURVE COEFFICIENTS

Dunlop Pneumatics: $K = a + b\Delta + c\Delta^2$

K = Kips per in. per meter effective length

	a	b	c	lower	upper
39	.176	-.025	.0023		
60	.0613	-.0092	.0009		
72	-1.725	+.286	-.0052		
90	.1149	-.0084	.0004		
108	.1807	-.0132	.0004		
126	-2.22	.2133	-.0023		
174	.0835	-.0038	.0001		

238

Dunlop Pneumatic Fenders

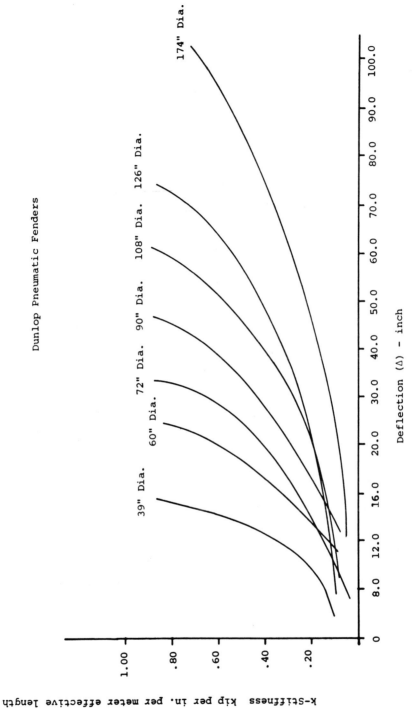

FIG.I-B: Stiffness vs. Deflection for Dunlop Pneumatic Fenders

APPENDIX I-C

GENERAL TIRE AND RUBBER COMPANY

Table I-C-1: GENERAL TIRE & RUBBER - RAYKIN
ENERGY TABLE

Kip-inch

	10	20	30	40	50	60	70	80	90	100	Δ_{max}
Size C											
20 ton	16	40	64	90	118	148	180	218	256	300	11.2
25	18	44	72	104	140	178	220	270	320	370	11.2
30	20	50	84	122	162	210	260	322	380	450	11.2
35	22	60	98	140	190	144	300	376	450	530	11.2
40	30	70	112	162	216	276	340	420	500	590	11.2
45	32	76	126	178	240	302	376	468	556	650	11.2
50	35	82	136	196	260	336	420	526	624	740	11.2
60	50	112	180	250	326	410	496	610	724	850	11.2
Size D											
20 ton		16	42	80	134	172	228	290	360	436	15.0
25		24	56	100	156	218	290	374	466	570	15.0
30		30	70	120	190	274	366	472	588	710	15.0
35		40	82	140	216	312	420	542	674	820	15.0
40		50	100	166	250	354	476	610	776	960	15.0
45		56	110	180	274	396	536	690	876	1064	15.0
50		70	130	210	316	450	606	780	980	1200	15.0
60		80	150	244	370	314	690	890	1120	1380	15.0
Size E											
20 ton		20	52	96	144	208	272	340	430	528	18.8
25		24	56	104	176	256	340	432	540	648	18.8
30		26	60	120	200	300	410	520	660	800	18.8
35		40	80	160	246	360	484	608	756	908	18.8
40		44	90	180	290	424	580	746	950	1180	18.8
45		54	116	210	340	484	660	850	1080	1330	18.8
50		56	120	230	366	534	720	936	1200	1520	18.8
60	30	84	170	290	444	620	830	1074	1380	1740	18.8
Size F											
20 ton		44	80	134	186	250	320	390	470	546	22.4
25		48	90	154	220	300	392	480	586	694	22.4
30		60	114	190	276	368	480	590	716	840	22.4
35		76	148	226	324	440	560	690	840	980	22.4
40		80	152	252	370	516	692	880	1100	1332	22.4
45		100	184	300	436	600	796	1010	1248	1500	22.4
50		116	214	340	490	670	884	1120	1380	1632	22.4
60	40	120	226	372	548	760	1020	1320	1660	2040	22.4

	10	20	30	40	50	60	70	80	90	100	Δ_{max}
Size G											
30 ton	28	74	136	200	290	380	496	620	750	890	26.1
35	32	88	160	248	348	456	584	736	880	1050	26.1
40	56	124	220	336	468	600	780	976	1170	1400	26.1
45	68	150	260	386	540	696	890	1104	1320	1570	26.1
50	80	176	300	450	620	800	1028	1280	1540	1830	26.1
60	90	200	350	520	720	940	1216	1540	1880	2280	26.1
Size H											
40	44	110	200	320	460	630	830	1056	1300	1560	30.0
45	60	148	264	416	600	800	1024	1270	1520	1790	30.0
50	78	176	300	470	660	900	1154	1436	1740	2050	30.0
60	82	194	340	520	748	1020	1310	1650	2020	2420	30.0
Size I											
50	124	166	420	610	820	1056	1310	1560	1956	2150	33.6
60	140	296	476	680	920	1200	1500	1800	2150	2530	33.6

Table I-C-2: GENERAL TIRE & RUBBER - RAYKIN
CURVE COEFFICIENTS

$$K = a + b\Delta \text{ or}$$
$$= a + b\Delta + c\Delta^2$$

Δ = inches
K = kips/inch

	a	b	c	boundary lower	boundary upper
Size C					
20 ton	5.66	-.026		0	11.2
25	7.26	-.066		0	11.2
30	9.10	-.090		0	11.2
35	10.50	-.092		0	11.2
40	11.94	-.056		0	11.2
45	13.38	-.064		0	11.2
50	13.96	-.020		0	11.2
60	16.80			0	11.2
Size D					
20 ton	4.38	-.026		0	15.0
25	5.50	-.038		0	15.0
30	7.08	-.072		0	15.0
35	8.10	-.074		0	15.0
40	9.20	-.066		0	15.0
45	10.26	-.072		0	15.0
50	11.08	-.054		0	15.0
60	12.86	-.054		0	15.0
Size E					
20 ton	3.76	-.066		0	18.8
25	4.64	-.078		0	18.8
30	5.90	-.100		0	18.8
35	6.76	-.110		0	18.8
40	7.36	-.054		0	18.8
45	8.26	-.056		0	18.8
50	9.64	-.076		0	18.8
60	10.60	-.076		0	18.8
Size F					
20 ton	4.511	-.147	.002	0	22.4
25	5.715	-.188	.003	0	22.4
30	5.691	-.214	.004	0	22.4
35	6.536	-.359	.004	0	22.4
40	7.978	-.303	.007	0	22.4
45	8.786	-.317	.007	0	22.4
50	9.438	-.351	.009	0	22.4
60	10.72	-.339	.009	0	22.4

	a	b	c	boundary lower	upper
Size G					
30 ton	4.96	-.114		0	26.1
35	5.42	-.114		0	26.1
40	5.48	-.080		0	26.1
45	6.14	-.086		0	26.1
50	6.70	-.066		0	26.1
60	7.52	-.044		0	26.1
Size H					
40 ton	5.32	-.088		0	30.0
45	6.18	-.106		0	30.0
50	6.36	-.088		0	30.0
60	7.10	-.086		0	30.0
Size I					
50 ton	5.78	-.082		0	33.6
60	6.92	-.098		0	33.6

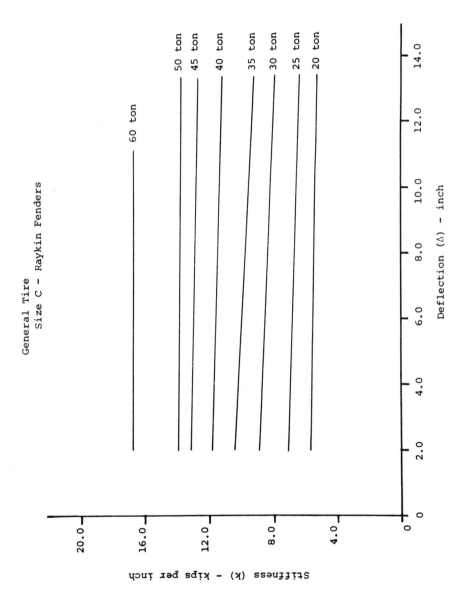

General Tire
Size C - Raykin Fenders

FIG. I-C-1: Stiffness vs. Deflection for General Tire Size C-Raykin Fenders

245

General Tire
Size D – Raykin Fenders

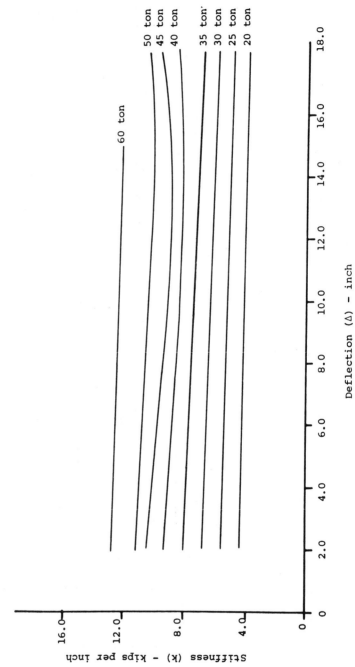

60 ton

50 ton
45 ton
40 ton

35 ton
30 ton
25 ton
20 ton

Deflection (Δ) – inch

Stiffness (k) – kips per inch

FIG. I-C-2: Stiffness vs. Deflection for General Tire Size D–Raykin Fenders

General Tire
Size E - Raykin Fender

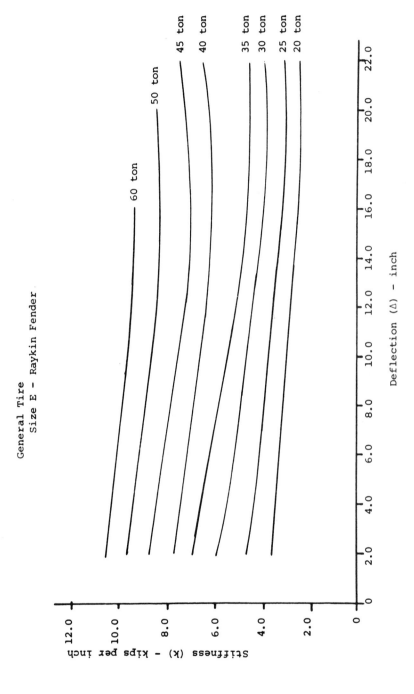

FIG. I-C-3: Stiffness vs. Deflection for General Tire Size E-Raykin Fenders

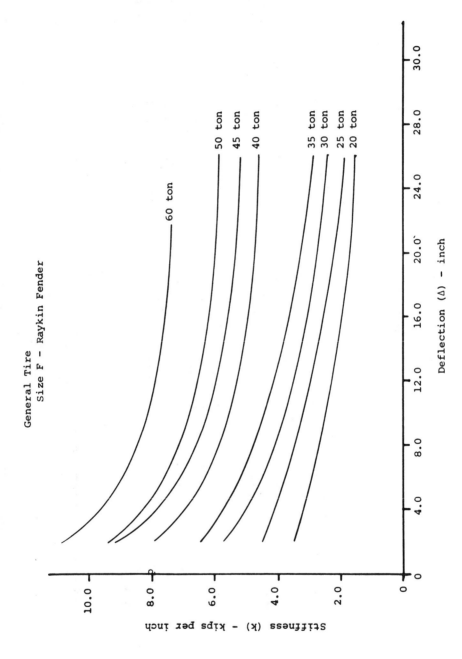

FIG. I-C-4: Stiffness vs. Deflection for General Tire Size F—Raykin Fenders

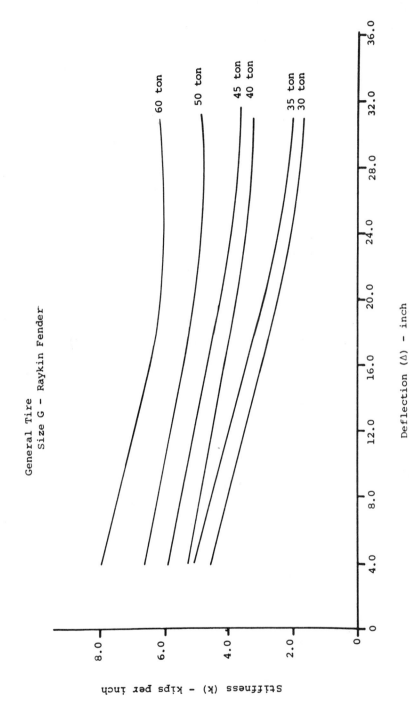

General Tire
Size G - Raykin Fender

Stiffness (k) - kips per inch

Deflection (Δ) - inch

FIG. I-C-5: Stiffness vs. Deflection for General Tire SizeG-Raykin Fenders

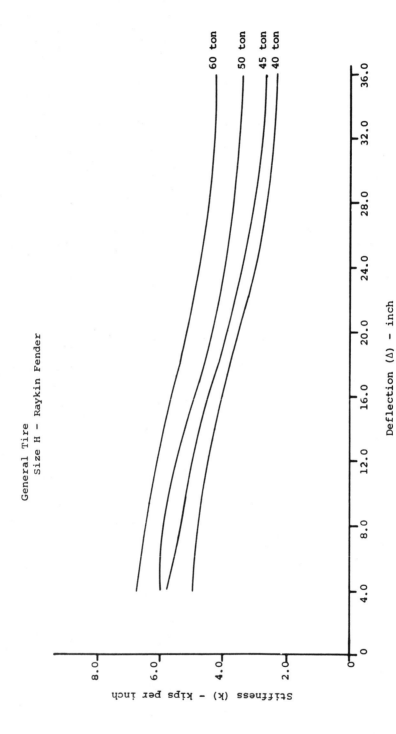

General Tire
Size H – Raykin Fender

60 ton
50 ton
45 ton
40 ton

Deflection (Δ) – inch

Stiffness (k) – kips per inch

FIG. I-C-6: Stiffness vs. Deflection for General Tire Size H-Raykin Fenders

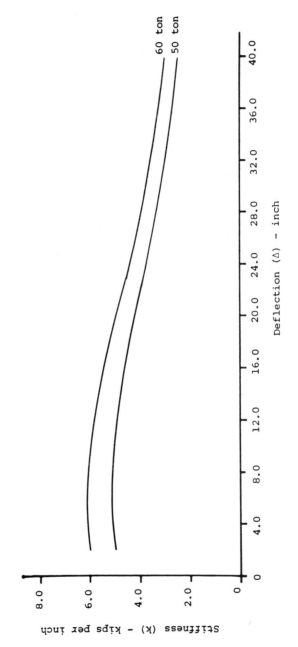

FIG. I-C-7: Stiffness vs. Deflection for General Tire Size I-Raykin Fenders

APPENDIX I-D

GOODYEAR TIRE AND RUBBER COMPANY

Table I-D-1: GOODYEAR ENERGY TABLES

K = Kip-in./foot of length

Rectangular	10	20	30	40	50	60	70	80	90	100	Δ_{max}
2x4x0								14		24	1.25
3.5x4.5x1					6.0	12	22	26	48	60	2.10
5x6.5x2.5			3.6	6.0		13	19	32	56	79	3.10
7x10x3		2.4	6.0	14.4	29	42	62	82	112	163	4.00
8x10x3		1.2	2.4	12.0	18	26	36	56	84	132	4.50
8x8x3		1.2	2.4	12.0	20	34	46	65	90	120	4.75
10x10x4		1.2	8.4	15.6	28	46	60	90	144	214	5.50
10x12x4		2.4	12.0	20.4	29	46	61	101	142	209	6.00
12x12x5		2.4	12.0	24.0	41	58	90	125	170	246	6.75
14x14x6		4.8	18.0	31.2	60	82	120	162	236	336	7.75
20x20x8		8.4	20.4	39.6	62	91	127	166	212	281	8.50

D-Fenders		Kip-in/foot of length									
6"		1.2	1.6	2.4	3.0	5.8	7.2	9.6	18	40	4.0
8"		1.2	2.4	3.6	6.0	9.0	13	17	28	53	5.5
12"		1.8	3.6	7.2	13.0	16	22	30	44	70	7.2

Trapezoid		Kip-in./foot of length									
10"	2.4	.6	8.4	23	16	20	28	32	37	48	4.6
13"	4.8	.7	12	17	26	30	40	54	66	72	6.0
15"	6.0	12	18	25	35	46	55	70	83	96	6.6
17"	6.0	14	24	34	47	60	77	91	110	127	7.6
20"	7.2	18	30	46	64	82	101	122	144	166	8.7

Cyl. End Ld.		Kip-in/foot of length			Δ_{max} = 50% of max. length						
10					24	36	48	72	96	120	7.5
12			12	24	60	84	96	144	180	216	9.0
15		24	36	60	96	144	204	252	324	396	11.5
18		36	60	108	168	252	336	432	528	648	13.2
21		48	96	168	276	396	528	1032	828	1020	15.5
24		60	120	204	300	408	540	696	864	1020	15.0
27		60	108	180	264	360	480	600	744	900	13.2

253

Cylindrical	10	20	30	40	50	60	70	Kip-in/foot length 80	90	100	Δ_{max}
3"							3.6	9.6	17	32	2.25
5"				2.4	2.4	3.6	4.8	6.0	12	28	3.5
7"			2.4	2.4	3.6	6.0	7.2	9.6	20	55	4.6
8"			2.4	3.6	6.0	8.4	9.6	16	31	59	5.4
10"		1.2	2.4	4.8	7.2	9.6	13	16	30	60	6.0
12"		2.4	3.6	6.0	9.6	13	17	22	36	72	7.0
15"		2.4	4.8	9.6	13	18	24	28	42	84	8.5
18"	1.2	3.6	7.2	12	18	26	34	41	59	108	10.5
21"	1.2	4.8	8.4	16	23	31	40	53	72	131	11.8
24"	1.2	6.0	11	18	26	36	48	62	80	140	13.0
27"	1.2	6.0	13	22	30	43	56	74.4	104	163	14.25
36"		1.2	24	48	60	84	108	156	216	300	21.0
48"		1.2	36	60	96	144	204	252	348	480	27.0
60"		2.4	48	84	132	192	264	360	480	648	31.0

Wing Type								Kip-in/foot of length			
3x1x6x7.5				2.4	3.6	4.8	6.0	9.0	11	19	1.75
4x1x6.5x1				2.4	4.8	7.2	8.4	12	22	26	2.10
4x2x6.5x1				3.6	4.8	6.0	9.6	16	28	57	2.50
6x2x9.5x1.5			3.6	7.2	9.6	13	19	28	36	60	3.3
6x3x9x1.5	1.2	3.6	4.8	6.0	7.2	9.6	13	22	37	84	4.0
10x3x16x2.5	1.2	3.6	7.2	12	17	24	32	43	64	96	4.7
10x4x16x2.5	1.2	3.6	7.2	12	18	24	32	43	61	94	5.0
8x4x12x2	1.2	3.6	6.0	7.2	8.4	12	16	22	38	67	4.6
12x6x18x3	2.4	4.8	7.2	9.6	13	18	23	32	50	94	6.5

Table I-D-2: GOODYEAR CURVE COEFFICIENTS

Rectangular $K = a + b + c^2$ or $= \text{constant} = d$ $K = \text{kips/inch/ft of length}$

	a	b	c	boundary lower	boundary upper	d	boundary lower	boundary upper
2x4x0	37.33	−52.00	58.67	.5	1.25			
3.5x4.5x1	35.50	−40.25	23.75	1.0	2.0			
5x6.5x2.5	35.19	−41.58	14.40	1.0	3.0			
7x10x3	29.64	−15.98	4.33	2.0	4.0			
8x10x3	19.36	−12.52	3.17	2.0	4.0			
10x10x4	13.08	− 6.13	1.54	2.0	5.0			
8x8x3	18.14	− 9.73	2.08	2.0	4.0			
10x12x4	52.00	−25.50	3.50	3.0	6.0	7.0	2.0	3.0
12x12x5	21.40	− 8.40	1.20	4.0	6.0	7.0	2.0	4.0
14x14x6	11.27	− 3.20	.533	4.0	7.0	7.0	2.0	4.0
20x20x8	20.68	− 5.34	.510	6.0	8.0	7.0	2.0	6.0

D Fenders $K = a + b\Delta + c\Delta^2$ $K = \text{kips/foot of length}$

	a	b	c	boundary lower	boundary upper
6 inch	41.25	−34.13	7.13	2.0	4.0
8 inch	7.100	− 4.73	.967	2.0	5.0
12 inch	3.58	− 1.76	.363	2.0	7.0

Regular and Wing Type Trapezoidal $K = a + b\Delta$ or $= a' + b'\Delta + c'\Delta^2$ $K/\text{foot of length}$

	a	b	boundary lower	boundary upper	a'	b'	c'	boundary lower	boundary upper
10"					7.53	− 1.56	.145	2.0	4.0
13"	4.66	+ .170	2.0	3.0	11.35	− 2.83	.255	3.0	5.0
15"	4.00	+ .500	2.0	3.0	10.75	− 2.32	.190	3.0	6.0
17"	4.00	+ .500	2.0	3.0	8.51	− 1.22	.074	3.0	7.0
20"	4.00	+ .500	2.0	4.0	10.49	− 1.36	.058	4.0	8.0

Wing Type Cylindrical $\quad$ K = a + bΔ $\quad$ or $\quad\quad$ K/foot of length

$$= a' + b'Δ + c'Δ^2$$

	a	b	boundary lower	boundary upper	a'	b'	c'	boundary lower	boundary upper
6x2x9.5	6.0	0	1.0	1.5	17.27	-13.87	4.26	1.5	3.0
10x3x16	6.0	0	1.0	2.5	10.00	- 3.43	.733	2.5	4.0
10x4x16	6.0	0	1.0	2.85	16.57	- 7.04	1.17	2.85	5.0
6x3x9	2.5	0	1.0	1.25	12.07	-12.98	4.10	1.25	3.0
8x4x12	2.5	0	1.0	3.25	25.90	-16.30	2.80	3.25	4.0
12x6x18	2.30	+.200	1.0	4.50	11.73	- 5.38	.764	4.50	6.0

Cylindrical $\quad\quad$ K = a + bΔ $\quad$ or $\quad\quad$ K/foot of length

$$= a' + b'Δ + c'Δ^2$$

	a	b	boundary lower	boundary upper	a'	b'	c'	boundary lower	boundary upper
5"					3.55	- 3.75	1.30	2.0	3.0
8"	1.20	0	2.0	3.2	33.51	-19.48	2.93	3.2	5.0
10"	1.20	0	2.0	4.2	79.25	-35.66	4.06	4.2	6.0
12"	1.20	0	2.0	5.0	45.55	-18.35	1.90	5.0	7.0
15"	1.20	0	2.0	6.9	120.12	-35.33	2.62	6.9	8.0
18"	1.20	0	2.0	8.2	24.27	- 6.30	.424	8.2	10.0
21"	1.20	0	2.0	9.5	99.27	-20.72	1.09	9.5	11.0
24"	1.20	0	2.0	11.0	107.59	-19.54	.898	11.0	13.0
27"	1.20	0	2.0	12.0	92.14	-15.36	.650	12.0	14.0
36"	1.20	0	2.0	15.0	12.31	- 1.50	.050	15.0	20.0
48"	1.20	0	2.0	16.0	- .404	.125	-.002	16.0	24.0
60"	1.20	0	2.0	20.0	2.99	- .191	.005	20.0	30.0

Cylindrical: End-Loaded $\quad\quad$ K = Constant d $\quad\quad$ K/foot of length

	d	boundary lower	boundary upper
10x5	3.60	0	7.0
12x6	4.60	0	9.0
15x7.5	5.90	0	11.0
18x9	7.18	0	13.0
21x10.5	8.40	0	15.0
24x12	9.28	0	15.0
27x13.5	10.50	0	13.0

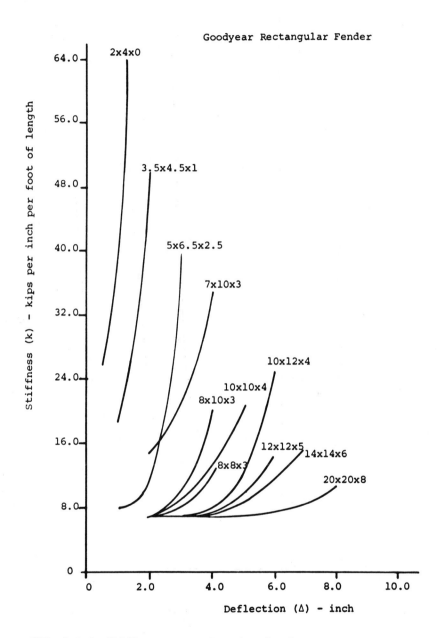

FIG. I-D-1: Stiffness vs. Deflection for Goodyear Rectangular Fenders

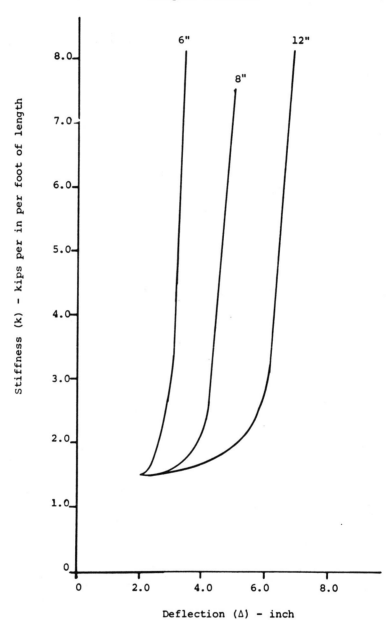

FIG. I-D-2: Stiffness vs. Deflection for Goodyear D-Fender

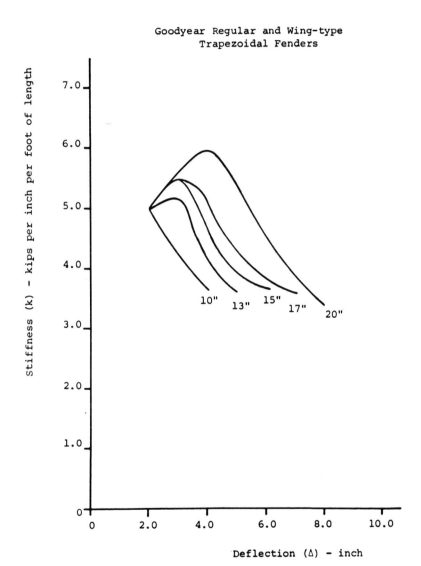

FIG. I-D-3: Stiffness vs. Deflection for Goodyear Regular and
Wing-type Trapezoidal Fenders

259

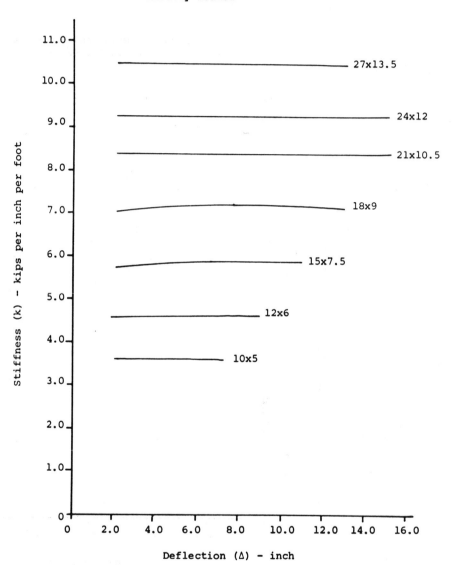

FIG. I-D-4: Stiffness vs. Deflection for Goodyear Cylindrical Fenders
            Axially Loaded

Goodyear Cylindrical Fenders

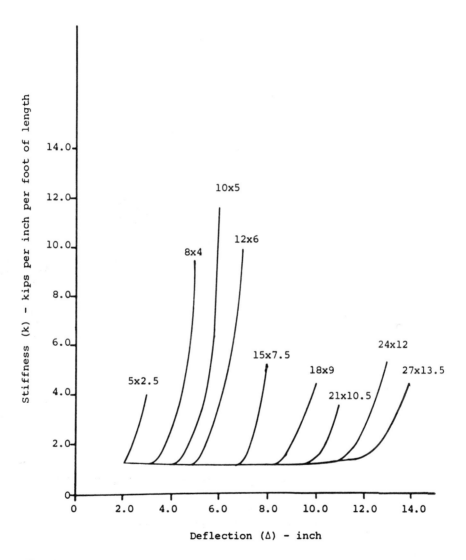

FIG. I-D-5: Stiffness vs. Deflection for Goodyear Cylindrical Fenders

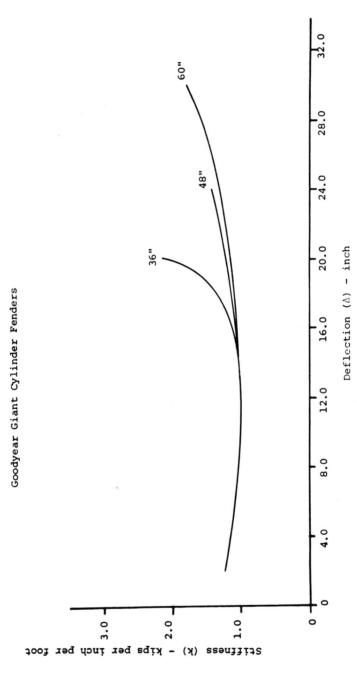

Goodyear Giant Cylinder Fenders

Stiffness (k) – kips per inch per foot

Deflection (Δ) – inch

FIG. I-D-6: Stiffness vs. Deflection for Goodyear Giant Cylindrical Fenders

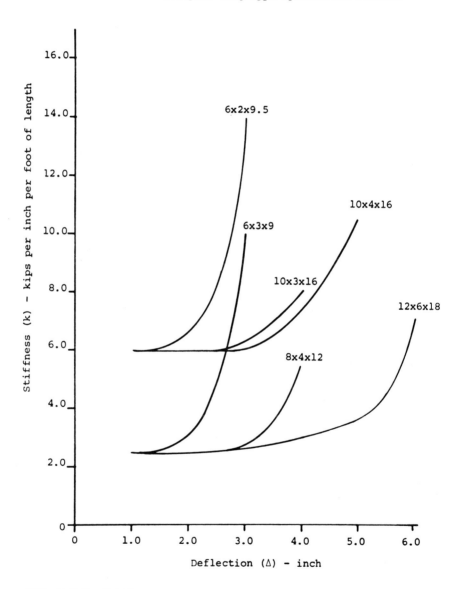

FIG. I-D-7: Stiffness vs. Deflection for Goodyear Wing-type
Cylindrical Fenders

Goodyear Portslide Fender

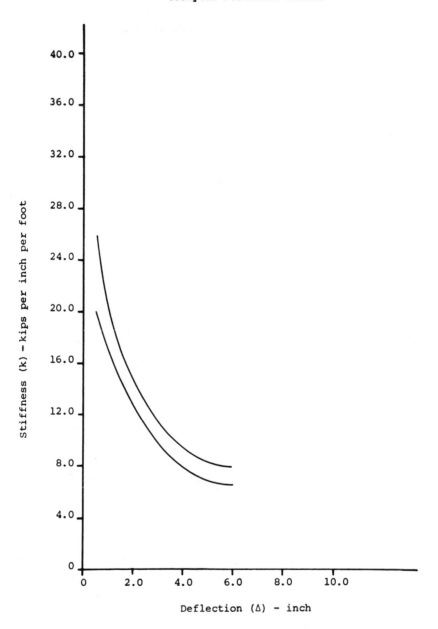

FIG. I-D-8: Stiffness vs. Deflection for Goodyear Portslide Fender

APPENDIX I-E

LORD CORPORATION

Table I-E-1: LORD ENERGY TABLE

	10	20	30	40	50	60	70	80	90	K = Kip-inches 100	Δ_{max}
1F4- 69			12	24	31	42	49	60	72	82	9.5
97			18	30	44	59	72	84	98	110	9.5
125			25	43	57	72	94	114	132	149	9.5
152			31	52	70	91	114	138	156	180	9.5
180			43	66	85	109	134	158	184	212	9.5
2F4-212				60	90	120	150	190	228	264	14.5
319				95	120	180	215	275	325	390	14.5
390				130	170	230	275	350	410	460	14.5
460				150	190	270	330	420	480	550	14.5
530				190	240	330	390	490	550	630	14.5
600				220	275	385	450	560	650	720	14.5
670				250	315	430	504	630	720	810	14.5
4F5- 14				3.0	5.4	7.2	9.6	12	15	18	5.5
21				5.5	7.8	11	14	18	22	26	5.5
28				7.2	10	14	19	23	28	33	5.5
35				9.0	13	19	23	29	36	42	5.5
47				12	18	24	31	38	48	58	5.5
5F- 600				200	275	360	455	540	635	720	23.0
765				240	360	445	565	685	810	890	23.0
900				300	444	540	660	820	960	1060	23.0
1070				360	515	660	816	960	1135	1260	23.0
1240				435	600	780	950	1105	1320	1450	23.0

Table I-E-2: LORD FENDER CURVE COEFFICIENTS

Lord Fender: $K = a + b\Delta + c\Delta^2$

1F4

	a	b	c	boundary lower	upper
69	3.79	- .431	.017	1.0	9.5
97	5.04	- .562	.023	1.0	9.5
125	6.89	- .934	.047	1.0	9.5
152	9.49	-1.60	.105	1.0	9.5
180	11.61	-1.69	.087	1.0	9.5

2F4

212	4.70	- .371	.011	2.0	14.5
319	6.94	- .553	.017	2.0	14.5
390	8.45	- .707	.023	2.0	14.5
460	10.16	- .776	.022	2.0	14.5
530	13.05	-1.08	.033	2.0	14.5
600	17.16	-1.69	.057	2.0	14.5
670	21.95	-2.40	.085	2.0	14.5

4F5

14	2.29	- .535	.050	1.0	5.0
21	3.13	- .515	.030	1.0	5.0
28	5.17	-1.35	.133	1.0	5.0
35	7.52	-2.27	.248	1.0	5.0
47	7.69	-2.00	.213	1.0	5.0

5F

600	5.70	- .390	.020	2.0	10.0
765	7.95	- .715	.040	2.0	10.0
900	10.25	- .978	.051	2.0	10.0
1070	13.42	-1.19	.048	2.0	10.0
1240	18.22	-1.74	.064	2.0	10.0

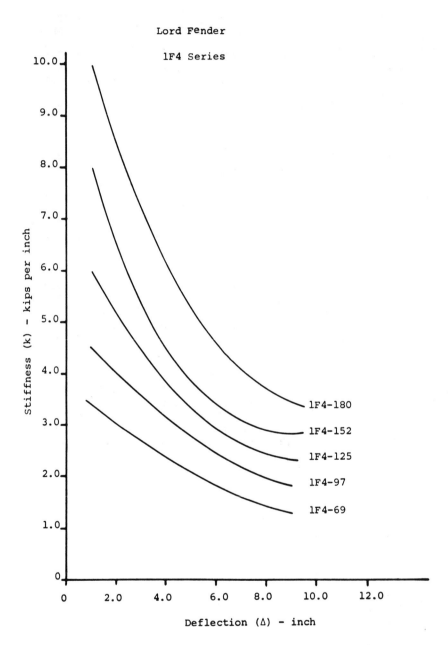

FIG. I-E-1: Stiffness vs. Deflection for Lord Fender 1F4 Series

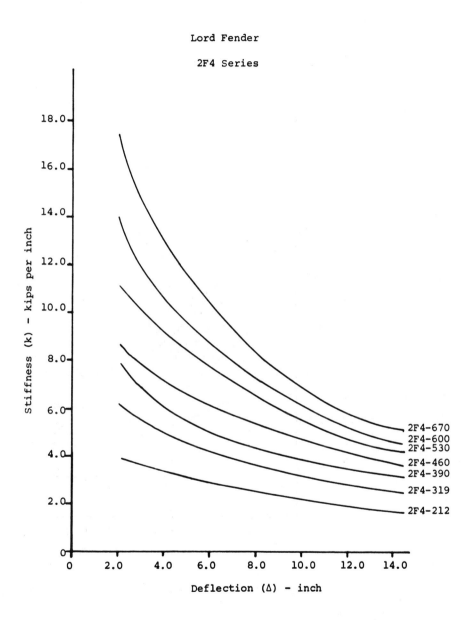

FIG. I-E-2: Stiffness vs. Deflection for Lord Fender 2F4 Series

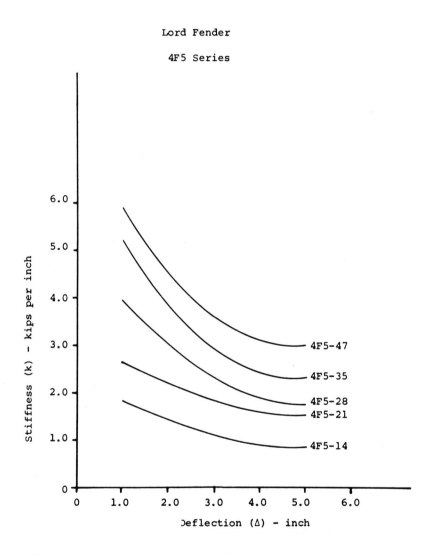

FIG. I-E-3: Stiffness vs. Deflection for Lord Fender 4F5 Series

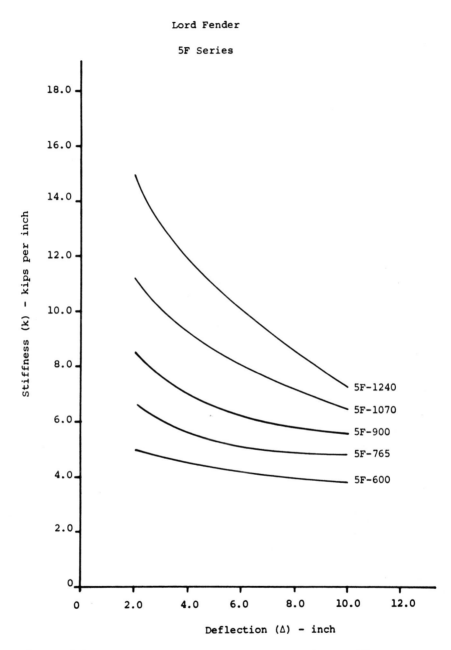

Lord Fender

5F Series

FIG. I-E-4: Stiffness vs. Deflection for Lord Fender 5F series

APPENDIX I-F

OIL STATES RUBBER COMPANY

Table I-F: OIL STATES
CURVE COEFFICIENTS AND ENERGY TABLE

Oil States Boat Bumpers $K = a + b\Delta + c\Delta^2$

			boundary	
a	b	c	lower	upper
12.88	-8.2	18.8	1.25	.25

							Kip-in./foot			
10	20	30	40	50	60	70	80	90	100	Δ_{max}
	.75	1.75	337	5.90	9.73	15.40	23.48	34.74	50	1.25

Oil States

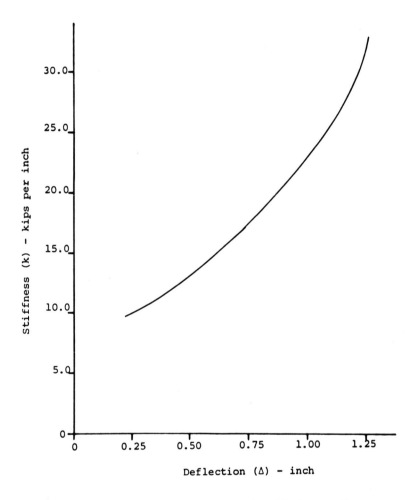

FIG. I-F: Stiffness vs. Deflection for Oil States Fender

APPENDIX I-G

SAMSON OCEAN SYSTEMS, INC.

Table I-G-1: SAMSON ENERGY TABLE

	10	20	30	40	50	E = Kip-inch 60
6x8	75	200	375	700	900	2040
6x12	120	280	520	1025	1800	3120
6x18	180	400	900	1600	2700	4680
8x12	200	400	960	2160	3180	5400
8x16	240	720	1440	2640	4200	7320
8x20	400	810	1700	3150	5100	9480
10x16	430	1030	2230	4000	6600	11500
10x20	360	960	2400	4320	7200	13200

Table I-G-2: SAMSON CURVE COEFFICIENTS

Solid Foam $K = a + b\Delta$ or
$\qquad\qquad\quad = a' + b'\Delta + c'\Delta^2$ $\qquad\qquad\qquad\qquad$ K = Kips per inch

			boundary					boundary	
	a	b	lower	upper	a'	b'	c'	lower	upper
10x20	.1478	.0003	24.0	56.0	.900	− .0262	.0002	56.0	72.0
4x8	---	---	--	--	.207	− .0049	.0002	7.0	28.5

Cost Fender $K = a + b\Delta$ or
$\qquad\qquad\quad = a' + b'\Delta + c'\Delta^2$

			boundary					boundary	
	a	b	lower	upper	a'	b'	c'	lower	upper
2'x3'									
32"x4'	.656	.0139	3.0	14.0	3.011	− .355	.0143	14.0	19.0
32"x6'	1.0	---	3.0	9.5	2.313	− .255	.0123	9.5	19.0
4'x6'	1.017	.0067	5.0	20.0	4.40	− .350	.0094	20.0	29.0
4'x8'	1.233	.0133	5.0	20.0	6.987	− .565	.0146	20.0	29.0
4'x10'	1.633	.0133	5.0	20.0	8.817	− .707	.0181	20.0	29.0

Sprayed Fender $K = a + b\Delta$ or
$\qquad\qquad\qquad = a' + b'\Delta + c'\Delta^2$

			boundary					boundary	
	a	b	lower	upper	a'	b'	c'	lower	upper
6x8	1.574	.018	7.0	32.0	16.150	− .904	.0146	32.0	43.0
6x12	2.670	.0114	7.0	29.0	5.925	− .301	.0068	29.0	43.0
6x18	3.878	.0174	7.0	30.0	17.38	− .920	.0162	30.0	43.0
8x12	2.467	.0133	9.0	40.0	28.317	−1.241	.0152	40.0	57.0
8x16	3.400	--	9.0	24.0	6.063	− .200	.0037	24.0	57.0
8x20	4.196	.0104	9.0	34.0	13.540	− .533	.0079	34.0	57.0
10x16	3.300	.0083	12.0	48.0	6.374	− .193	.0029	48.0	72.0
10x20	4.300	.0083	12.0	48.0	12.76	− .403	.0049	48.0	72.0

Samson Solid Foam Fenders

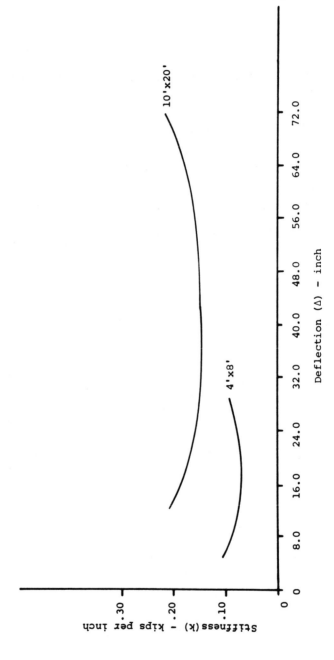

FIG. I-G-1: Stiffness vs. Deflection for Samson Solid Foam Fenders

Samson Cast Fender

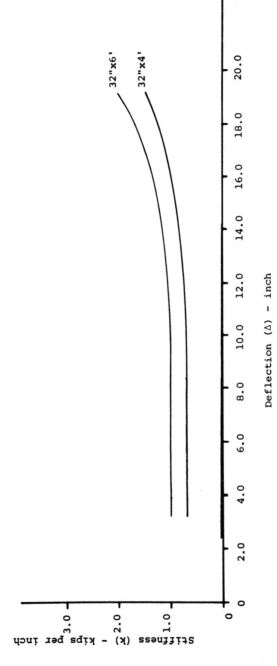

FIG. I-G-2: Stiffness vs. Deflection for Samson Cast Fenders up to 32" x 6'

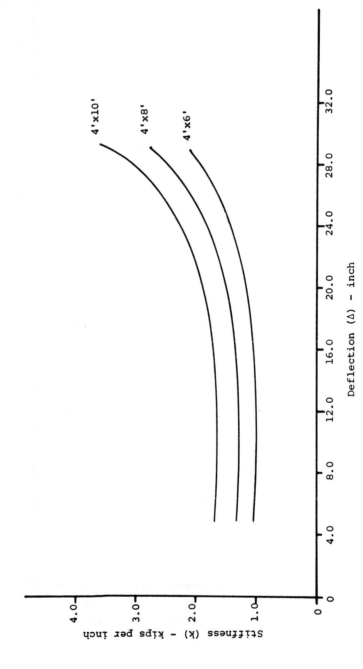

Samson Cast Fender

Stiffness (k) - kips per inch

Deflection (Δ) - inch

4'x10'

4'x8'

4'x6'

FIG. I-G-3: Stiffness vs. Deflection for Samson Cast Fenders up to 4' x 10'

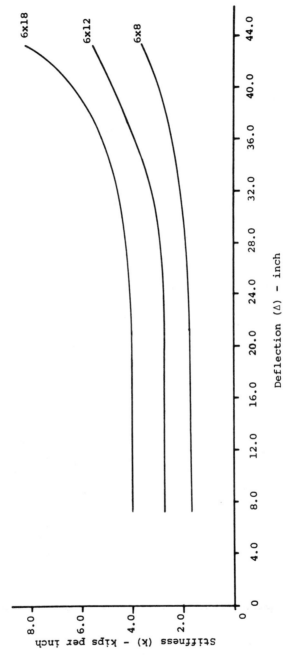

Samson Sprayed Fender

FIG. I-G-4: Stiffness vs. Deflection for Samson Sprayed Fender up to 6' x 18'

Samson Sprayed Fender

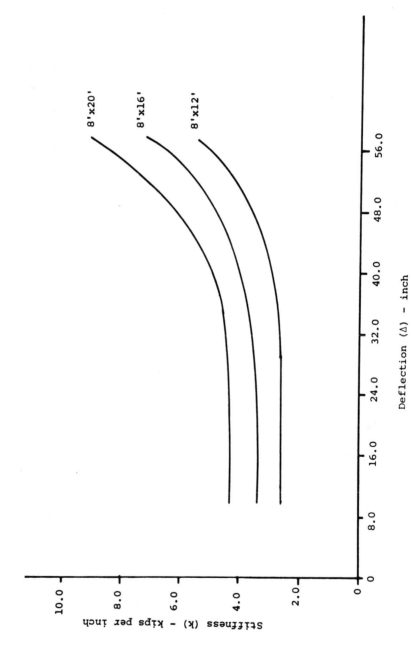

FIG. I-G-5: Stiffness vs. Deflection for Samson Sprayed Fender up to 8' x 20'

Samson Sprayed Fender

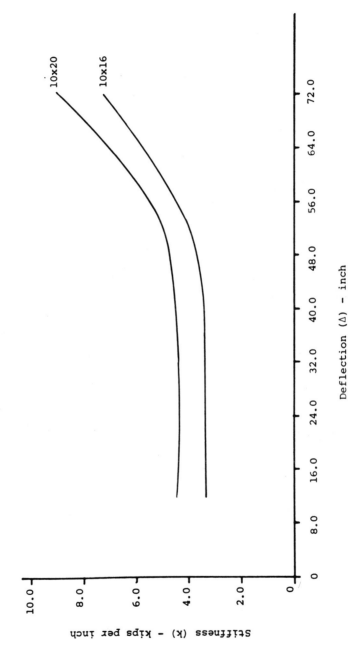

FIG. I-G-6: Stiffness vs. Deflection for Samson Sprayed Fender up to 10'x20'

APPENDIX I-H

SEWARD INTERNATIONAL, INC.

Table I-H: SEWARD
CURVE COEFFICIENTS AND ENERGY TABLE

Sea Cushion $K = a + b\Delta$ or   K = Kips per inch
$\quad\quad\quad\quad = a' + b'\Delta + c'\Delta^2$

	a	b	boundary lower	upper	a'	b'	c'	boundary lower	upper
3x5	.912	+.0095	3.5	14.0	2.88	−.241	.0082	14.0	21.6
4x7.4	1.427	+.0071	4.0	18.0	3.52	−.209	.0057	18.0	34.0
5x10	1.866	+.0056	6.0	24.0	3.61	−.177	.0045	24.0	42.0
6x12	1.926	+.0217	8.0	31.0	3.44	−.120	.0029	31.0	50.0
8x16	2.882	+.0218	10.0	44.0	10.01	−.308	.0038	44.0	65.0
10x16	1.918	+.0235	12.0	44.0	6.47	−.163	.0020	44.0	84.0
10x20	2.267	+.0361	12.0	48.0	7.50	−.187	.0024	48.0	84.0
11x22	3.900	+.0167	12.0	66.0	31.20	−.788	.0059	66.0	92.0

E = Kip-inch

	10	20	30	40	50	60	70	80	90	100	Δ_{max}
3x5		25	60	105	168	265	405				36"
4x7.4		60	144	260	435	690	1080				48"
5x10		120	300	690	900	1470	2280				60"
6x12		210	510	975	1560	2520	3960				72"
8x16		600	1320	2280	3960	6000	9300				96"
10x16		700	1960	3600	6900	9240					120"
10x20		960	2400	4500	7500	11400	18000				120"
11x22		1200	3300	6000	9900	15300	24000				132"

Seward Sea-Cushion Fender

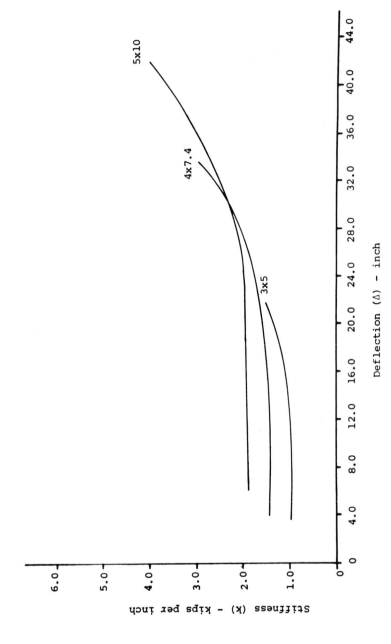

FIG. I-H-1: Stiffness vs. Deflection for Seward Sea Cushion Fender up to 5' x 10'

Seward Sea-Cushion Fender

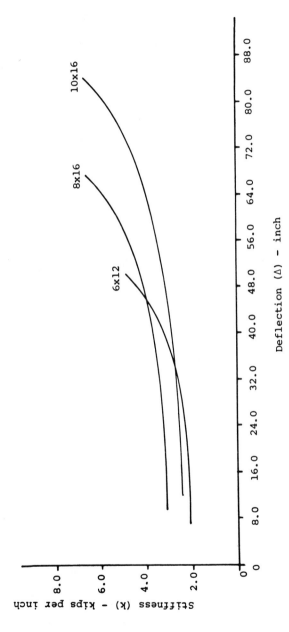

FIG. I-H-2: Stiffness vs. Deflection for Seward Sea Cushion Fender up to 10' x 16'

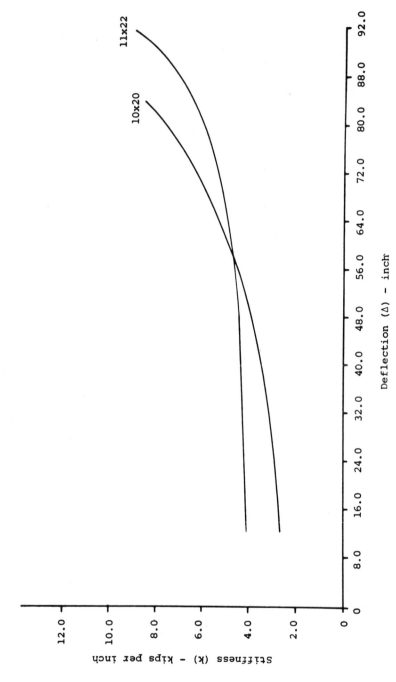

Seward Sea-Cushion Fender

FIG. I-H-3: Stiffness vs. Deflection for Seward Sea Cushion Fender up to 11' x 22'

APPENDIX I-J

UNIROYAL, INC.

Table I-J-1: UNIROYAL ENERGY TABLE

Delta	Δ_{max} = inches					K = kip-in/foot of length					Δ_{max}
	10	20	30	40	50	60	70	80	90	100	
13A	2.4	4.8	11	16	24	34	34	55	65	78	5.7
20A	3.6	9.6	16	24	37	52	67	83	101	116	7.3
23A	4.8	13	28	46	66	90	113	137	162	185	9.5
28A	7.2	19	37	62	91	120	154	187	221	252	11.0
10	2.4	4.8	7.2	11	16	20	25	31	36	41	4.5
15	3.6	6.0	12	18	26	35	46	56	67	79	5.9
20	4.8	11	20	31	44	60	76	92	108	125	7.8
25	7.2	18	32	50	72	97	120	146	172	199	9.9

Cylindrical

Delta	10	20	30	40	50	60	70	80	90	100	Δ_{max}
5			6.0	9.6	11	12	18	30	48	72	4.3
7			8.4	12	13	14	16	30	48	96	5.3
8			9.6	11	12	14	18	36	60	106	6.0
10		7.2	11	12	16	22	24	48	84	168	7.3
12		8.4	12	14	22	25	36	72	102	228	9.0
15	4.8	12	14	22	28	36	48	84	156	336	11.1
18	6.0	12	18	24	36	50	66	108	180	414	13.3
21	6.0	12	22	32	48	62	84	144	228	492	15.3
24	9.6	16	28	42	60	84	128	252	312	612	18.0
28		12	16	30	46	76	104	134	168	228	17.0
32		12	24	36	60	90	132	168	210	288	19.0
36		12	24	54	76	112	150	204	254	348	21.0
48		16	46	84	136	196	264	344	426	576	27.5
60	12	30	68	120	196	277	376	496	630	828	33.0

Rectangular

Delta	10	20	30	40	50	60	70	80	90	100	Δ_{max}
3.5x4.5x1	3.0	8.4	12	18	24	31	38	49	61	96	2.3
5x6x2.5	3.0	6.0	11	14	23	30	44	58	78	100	3.3
5x6x2.5	3.6	7.2	12	17	24	30	37	48	80	122	3.6
6x6.5x2.5	4.8	9.6	13	22	28	38	49	64	102	198	4.1
8x10x3	4.8	11	17	28	38	52	70	92	131	216	5.0
8x8x3	4.8	12	22	26	36	52	73	103	146	240	5.5
10x12x4	6.0	16	24	32	50	72	98	142	209	336	6.75
12x12x5	7.2	20	30	36	62	188	125	181	250	384	8.0
14x14x6	11	23	35	50	72	102	140	197	264	384	9.0

										E = Kip-in./foot of length	
Cyl. End Ld.	10	20	30	40	50	60	70	80	90	100	Δ_{max}
10	12	18	28	36	46	64	84	90	126	172	7.8
12	18	28	36	46	64	90	126	180	216	270	9.8
15	36	54	100	144	198	244	298	360	450	540	12.8
18	46	90	144	198	270	342	414	504	604	784	14.5
21	82	154	234	342	450	558	702	850	1026	1260	17.8
24	100	190	316	456	612	792	980	1188	1458	1800	19.7

Wing Type

	10	20	30	40	50	60	70	80	90	100	Δ_{max}
6x2				3.0	7.2	17	26	37	58	84	3.7
6x3					6.0	13	24	38	59	95	4.1
8x4				4.8	14	30	52	84	120	178	5.6
10x4				9.6	23	38	66	73	148	216	6.8
12x5			6.0	16	34	59	89	137	216	562	8.3

Table I-J-2: UNIROYAL CURVE COEFFICIENTS

Wing Type $K = a + b\Delta + c\Delta^2$ or K = kips/inch/foot of length
                 $= a' + b'\Delta$ Δ = inch

	a	b	c	boundary lower	boundary upper	a'	b'	boundary lower	boundary upper
6x2	14.03	-13.98	4.95	1.0	3.5	---	---	---	---
6x3	12.90	-14.42	4.52	1.0	4.0	---	---	---	---
8x4	3.60	- 4.21	1.61	1.0	5.5	---	---	---	---
10x4	3.49	- 3.56	1.07	3.0	6.75	- .80	+ 1.8	1.0	3.0
12x5	2.50	- 2.07	.561	3.0	8.25	0	+ 1.00	1.0	3.0

Cylindrical $K = a + b\Delta$ or
                   $= a' + b'\Delta + c'\Delta^2$

	a	b	boundary lower	boundary upper	a'	b'	c'	boundary lower	boundary upper
5"	1.82	- .018	1.0	2.2	9.34	- 12.90	4.19	2.2	4.1
7"	1.82	- .018	1.0	3.3	19.18	- 15.56	3.11	3.3	5.25
8"	1.82	- .018	1.0	3.8	6.46	- 9.28	2.07	3.8	6.00
10"	1.82	- .018	1.0	5.0	80.15	- 33.37	3.53	5.0	7.40
12"	1.82	- .018	1.0	5.9	25.10	- 10.32	1.08	5.9	9.00
15"	1.82	- .018	1.0	8.5	133.20	- 33.23	2.09	8.5	11.25
18"	1.82	- .018	1.0	9.0	60.74	- 13.32	.750	9.0	13.40
21"	1.82	- .018	1.0	10.5	55.92	- 10.41	.500	10.5	15.75
24"	1.82	- .018	1.0	12.0	47.00	- 7.83	.336	12.0	17.90
28"	1.24	+ .004	2.0	13.0	19.21	- 2.70	.101	13.0	16.50
32"	1.24	+ .004	2.0	13.5	33.95	- 4.15	.132	13.5	18.50
36"	1.24	+ .004	2.0	17.0	52.63	- 5.71	.158	17.0	21.30
48"	1.24	+ .004	2.0	23.5	69.36	- 5.07	.119	23.5	28.0
60"	1.24	+ .004	2.0	27.0	52.13	- 3.49	.060	27.0	35.50

Rectangular $K = a + b\Delta$ or
                    $= a' + b'\Delta + c'\Delta^2$

	a	b	boundary lower	boundary upper	a'	b'	c'	boundary lower	boundary upper
3.5x4.5x1	23.0	- .50	.50	1.0	38.72	- 46.08	23.36	1.0	2.25
5x6x2.5	23.0	- .50	.50	1.50	176.56	-145.92	32.32	1.5	3.25
5x6x2.5	23.0	- .50	.50	2.0	33.03	- 22.57	5.77	2.0	3.75
6x6.5x2.5	23.0	- .50	.50	2.0	44.57	- 28.05	5.63	2.0	4.00
8x10x3	23.0	- .50	.50	2.0	69.50	- 38.03	6.267	2.0	5.0
8x8x3	10.84	- 1.67	.5	2.0	40.60	- 21.70	3.700	2.0	5.0
10x12x4	10.84	- 1.67	.5	2.5	17.27	- 7.40	1.35	2.5	6.5
12x12x5	10.84	- 1.67	.5	3.0	17.88	- 6.70	.988	3.0	8.0
14x14x6	10.84	- 1.67	.5	3.5	15.11	- 4.64	.632	3.5	8.75

Delta and Delta A   K = a + bΔ

	a	b	boundary lower	upper
13A	7.00	- .636	0.0	6.0
20A	7.28	- .600	0.0	7.5
23A	7.75	- .581	0.0	8.5
28A	8.00	- .528	0.0	10.0
10	7.70	-1.20	1.0	4.0
15	6.65	- .580	0.0	5.5
20	7.20	- .579	0.0	7.5
25	8.09	- .588	0.0	9.0

Cylindrical End-Loaded K = Constant or
                          = a + bΔ

	a	b	boundary lower	upper
10x5	3.85	- .075	2.0	8.0
12x6	4.5	---	2.0	10.0
15x75	6.0	---	2.0	13.0
18x9	7.0	---	4.0	14.0
21x10.5	8.55	- .025	2.0	18.0
24x12	9.3	---	4.0	20.0

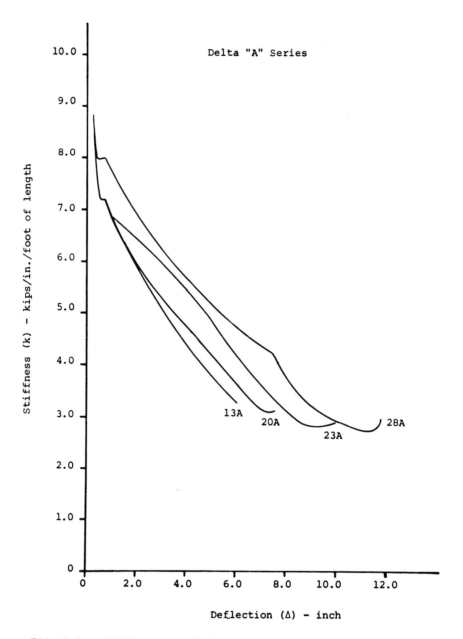

FIG. I-J-1: Stiffness vs. Deflection for Uniroyal Delta A Series

Delta Series

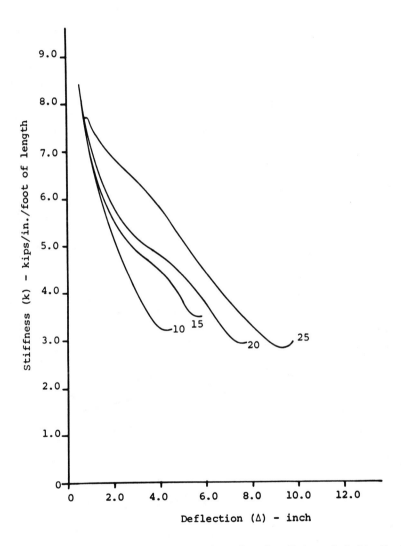

FIG. I-J-2: Stiffness vs. Deflection for Uniroyal Delta Series

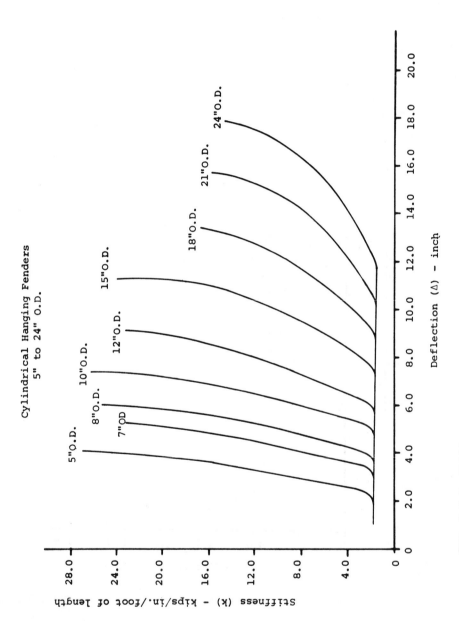

Cylindrical Hanging Fenders
5" to 24" O.D.

FIG. I-J-3: Stiffness vs. Deflection for Uniroyal Cylindrical Fenders up to 24" O.D.

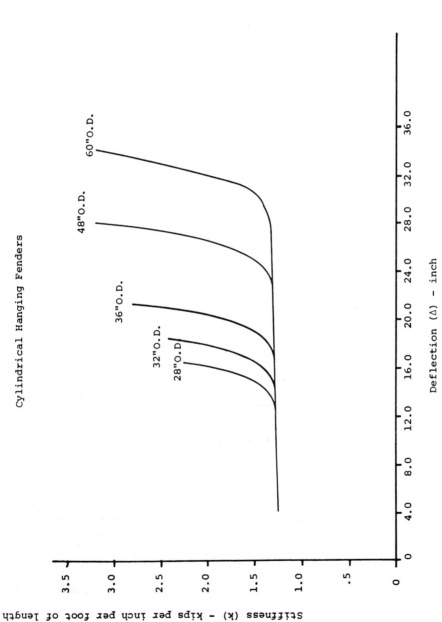

Cylindrical Hanging Fenders

FIG. I-J-4: Stiffness vs. Deflection for Uniroyal Cylindrical Hanging Fenders up to 60" O.D.

297

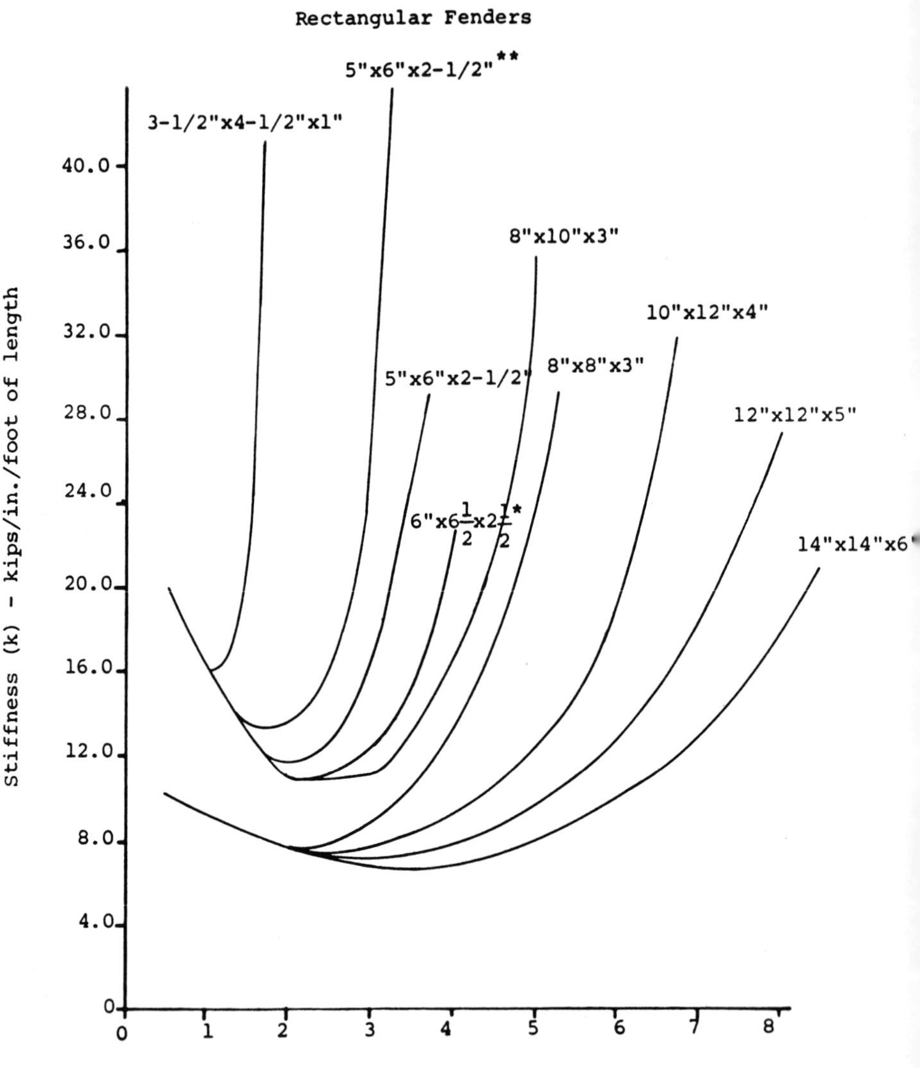

FIG. I-J-5: Stiffness vs. Deflection for Uniroyal Rectangular Fenders
Axially Loaded

Rectangular Fenders - Shear Load

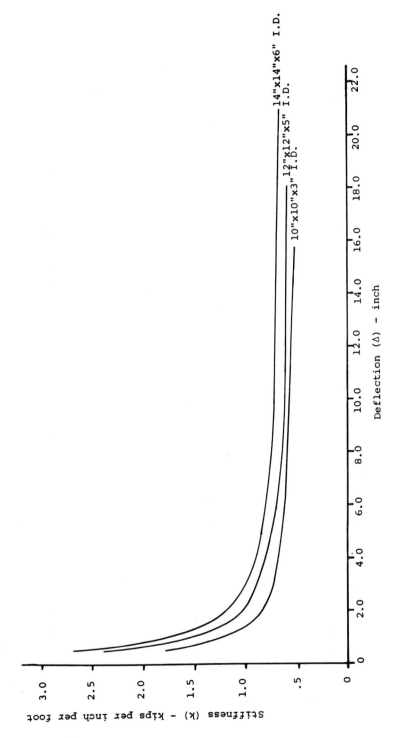

FIG. I-J-5: Stiffness vs. Deflection for Uniroyal Rectangular Fenders in Shear

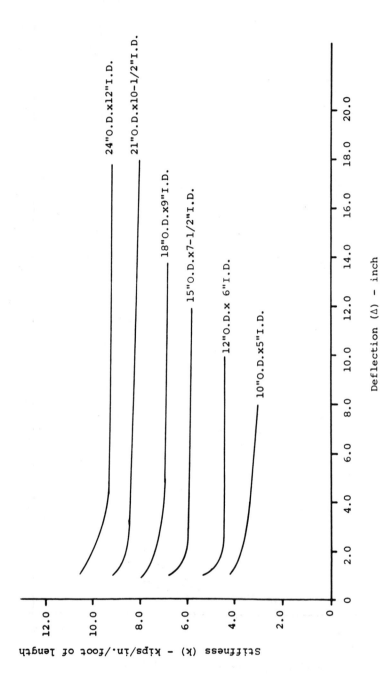

Cylindrical Fenders   End loaded

24"O.D.x12"I.D.

21"O.D.x10-1/2"I.D.

18"O.D.x9"I.D.

15"O.D.x7-1/2"I.D.

12"O.D.x 6"I.D.

10"O.D.x5"I.D.

Deflection (Δ) - inch

Stiffness (k) - kips/in./foot of length

FIG. I-J-7: Stiffness vs. Deflection for Uniroyal Cylindrical Fenders End Loaded

300

Wing-type Fenders

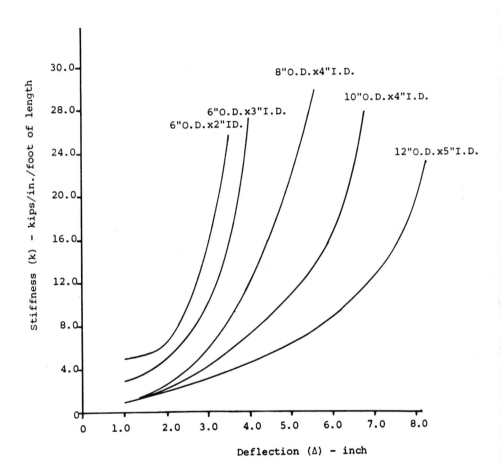

FIG. I-J-8: Stiffness vs. Deflection for Uniroyal Wing-type Fenders

APPENDIX I-K

YOKOHAMA RUBBER COMPANY, LTD.

Table I-K-1: YOKOHAMA ENERGY TABLE

Pneumatics   E = Kip-in.

	10	20	30	40	50	60	70	80	90	100	Δ_{max}
700x1500		6.0	15	39	72	132					27.6
1000x1500		11	25	60	120	240					39.4
1000x2000	3	15	39	85	174	350					39.4
1350x2500	12	30	84	195	390	780					53.1
1500x3000	18	54	120	290	610	1150					59.1
1700x3000	20	72	180	370	750	1440					66.9
2000x3500	24	105	240	550	1170	2315					78.7
2500x5500		360	900	1800	3900	6900					98.4
3300x6500		720	2040	4560	8940	16200					129.9
4500x9000		2100	5400	12000	23750	42300					177.2

Air Block   E = Kip-in.

	10	20	30	40	50	60	70	80	90	100	Δ_{max}
300x 400		2.2	5.2	9.6	17	26	41				
450x 600		8.7	17	35	52	96	143				
600x 800		17	39	78	130	210	320				
750x1000		43	87	156	260	420	650				
900x1200		80	150	280	435	715	1112				
1200x1600		175	350	650	1043	1650	2700				
1500x2000		300	735	1260	2085	3475	5213				

Table I-K-2: YOKOHAMA CURVE COEFFICIENTS

Pneumatics: $K = a + b\Delta + c\Delta^2$

K = Kips per inc
Δ = inch

	a	b	c	boundary lower	boundary upper
700x1500	.397	-.032	.0076	1.38	15.16
1000x1500	.346	-.026	.0050	1.97	23.67
1000x2000	1.317	-.150	.0072	2.00	23.62
1350x2500	.552	-.041	.0037	3.94	31.50
1500x3000	.748	-.040	.0034	7.87	35.43
1700x3000	.633	-.039	.0030	3.94	38.00
2000x3500	.967	-.047	.0023	4.00	47.20
2500x5500	1.414	-.017	.0019	7.87	55.12
3300x6500	2.170	-.032	.0016	15.75	78.74
4500x9000	2.734	-.018	.0011	19.69	98.43

Air Blocks: $K = a + b\Delta + c\Delta^2$

	a	b	c	lower	upper
300x 400	.820	-.099	.028	1.0	7.90
450x 600	1.474	-.163	.023	1.95	11.80
600x 800	1.793	-.140	.017	2.30	16.55
750x1000	4.493	-.413	.020	3.95	19.70
900x1200	2.837	-.099	.0086	4.00	23.60
1500x2000	5.894	-.200	.0066	8.00	39.40
1200x1600	5.608	-.290	.0103	4.0	31.50

Yokohama Pneumatic Fender

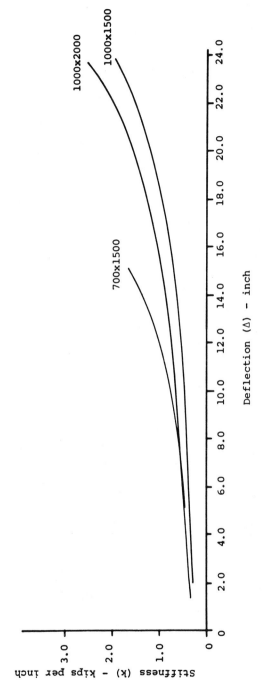

FIG. I-K-1: Stiffness vs. Deflection for Yokohama Pneumatic Fender up to 1000 x 2000

Yokohama Pneumatic Fenders

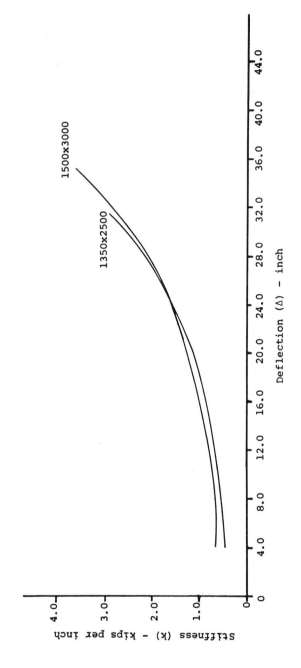

FIG. I-K-2: Stiffness vs. Deflection for Yokohama Pneumatic Fenders up to 1500 x 3000

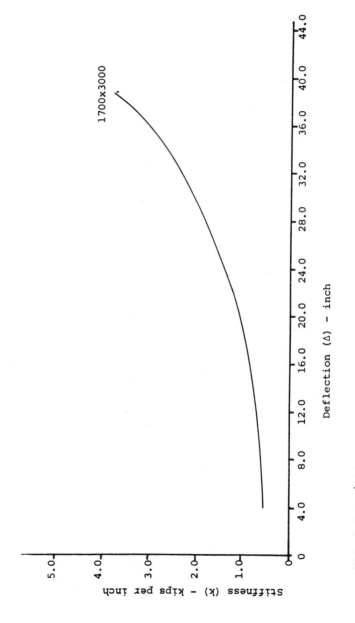

Yokohama Pneumatic Fender

FIG. I-K-3: Stiffness vs. Deflection for Yokohama Pneumatic Fender 1700 x 3000

Yokohama Pneumatic Fender

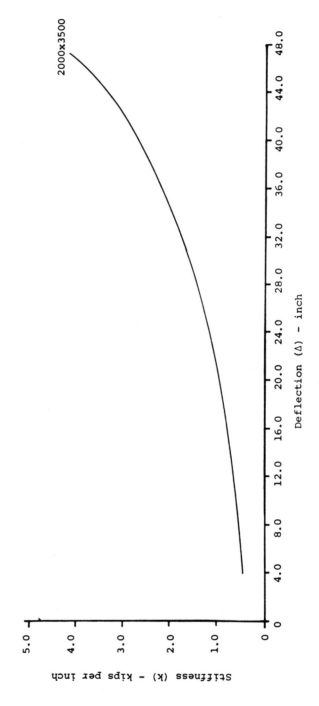

FIG. I-K-4: Stiffness vs. Deflection for Yokohama Pneumatic Fender 2000 x 3500

Yokohama Pneumatic Fender

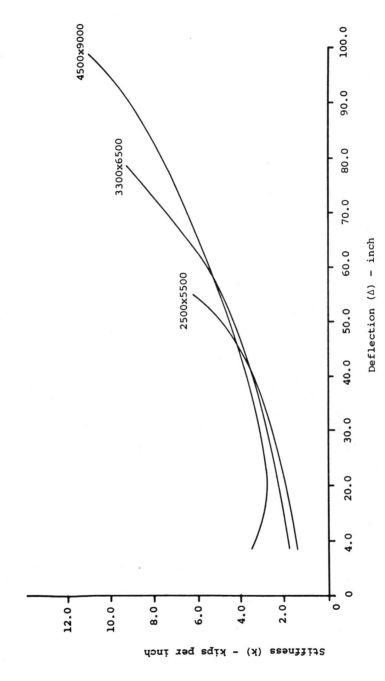

FIG. I-K-5: Stiffness vs. Deflection for Yokohama Pneumatic Fender up to 4500 x 9000

Yokohama Air Block Fenders

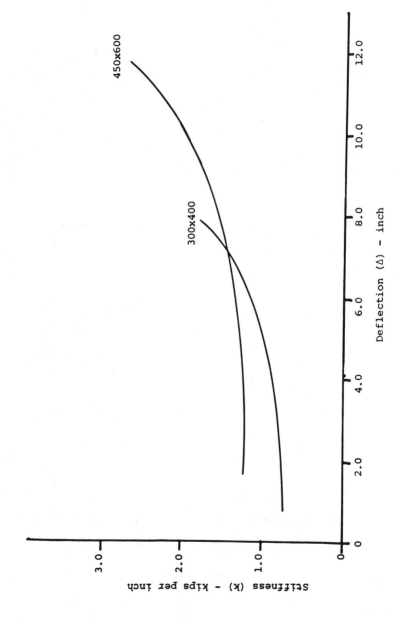

FIG. I-K-6: Stiffness vs. Deflection for Yokohama Air Blocks up to 450 x 600

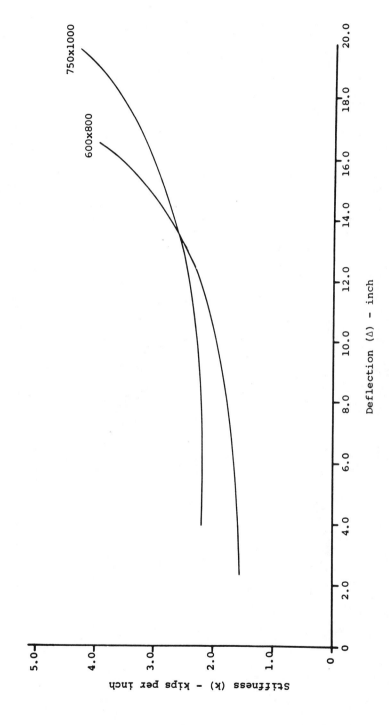

Yokohama Air Block Fenders

750x1000

600x800

Deflection (Δ) - inch

Stiffness (k) - kips per inch

FIG. I-K-7: Stiffness vs. Deflection for Yokohama Air Blocks up to 750 x 1000

Yokohama Air Block Fenders

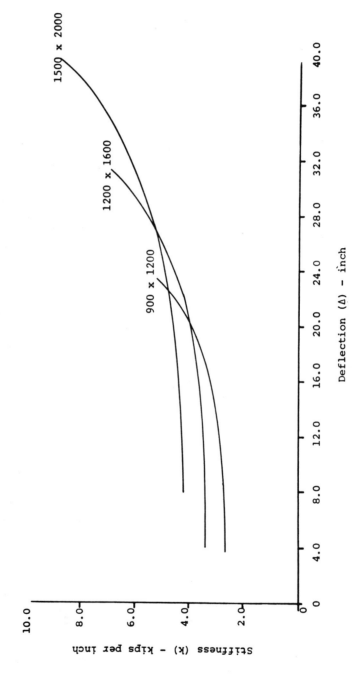

FIG. I-K-8: Stiffness vs. Deflection for Yokohama Air Blocks up to 1500 x 2000

APPENDIX II

REFERENCES OF COMPUTER WORK

1. State-of-the-Art - Bridge Protective Devices, United States Coast Guard Contract CG-71955-A.

2. A. DeQuinn, "Design and Construction of Ports and Marine Structures," McGraw Hill Book Co., N.Y., 1972.

3. C. P. Heins, Applied Plate Theory for the Engineer, Lexington Books, Lexington, Mass., 1976.

APPENDIX III

NOTATIONS

M = mass of ship (W/g)

W = weight of ship

g = gravity (32.2 ft/sec^2)

Δ_s = deformation of fender systems at point of impact

$v_i v_f$ = initial and final velocity of ship

E = modulus of elasticity

I = moment of inertia of pile or waler

L = cantilever height of pile

k = spring constant of fender

$D.F.$ = load distribution factor

M = induced pile moment

F_a = applied force to pile due to ship

F_r = resisting pile force

S = section modulus of pile

C_H = hydrodynamic coefficient

C_E = eccentricity coefficient of ship and fendering

C_s = softness coefficient of ship

C_c = configuration coefficient of ship

E_{in} = input energy due to ship

E_o = output energy available from pile

Δ_p = single pile deformation

Δ_f = fender deformation

F = applied force to pile due to ship

q, q_x, q_y = total external uniform load, and load on waler and pile respectively

w = lateral deformation of grid

314

x = horizontal coordinate

y = vertical coordinate

D_x = grid stiffness in x direction = $EI_x/\lambda y$

D_y = grid stiffness in y direction = $EI_y/\lambda x$

$\sigma = D_x/D_y$

$n = \lambda x/\lambda y$

REFERENCES

1. Public Works Maintenance-Wharfbuilding; "Structural Timbers-Fender Systems", U.S. Navy Wharfbuilding Handbook 1961; pp. 276-280.

2. Gower, L., Mardel, C. and Heathorn, W.; "The Behavior of Piles in Maritime Works", Insititute of Civil Engineers: Behavior of Piles; Paper 14, pp. 153-159.

3. Amer. Assoc. of Port Authorities; Port Design and Construction; pp1 140-152, 108-119.

4. Morison, J.R., Johnson, J.W., and O'Brien, M.P.; "Experimental Studies of Forces on Piles"; Coastal Engineering; pp. 340-369.

5. Reese, Lymon C. and Haliburton, T.A.; "Dynamic Response and Stability of Piers and Piles"; ASCE Proc. (Waterways 2) Vol. 89; May 1963, pp. 85-89.

6. Michalos, James; "Dynamic Response and Stability of Piers and Piles"; ASCE Proc. (Waterways 3) Vo. 88; August 1962; pp. 61-83.

7. Wakeman, C.M., Dockweiler, E.V., Stover, H.E. and Whiteneck. L.L.; "Use of Concrete in Marine Environments"; Journal of the American Concrete Institute; April 1958; Vol. 29; pp. 841-856.

8. Seiler, J.F. and Keeney, W. D.; "The Efficiency of Piles in Groups"; Wood Preserving News; November 1944; pp.109-118.

9. Praeger, E.H.; "Energy Absorption by the Pier and Wharf Structure"; Princeton Conference on Berthing in Exposed Locations.

10. Kleiner; "Construction of Dolphins in the Martinsa Bay"; Papers of 5th International Harbor Congress.

11. Volse, Louis A.; "Docking Fenders to Pier Protection"; Engineering News-Record; Vol. 160, No. 19; May 15, 1958; pp. 40-44.

12. Weis, John and Blancato, F.; "A Breasting Dolphin for Berthing Super Tankers"; ASCE Proc. (Waterways 3), September 1959; pp. 183-195.

13. Shu-t'len Li, F. and Venkataswamy Ramakrishnan, M.; "Ultimate Energy Design of Prestressed Concrete Fender Piling"; ASCE Proc. (Waterways 4) Vol. 97; November 1971; pp. 647-662.

14. Shu-t'len Li, F.; "Operative Energy Concept in Marine Fendering"; ASCE Proc. (Waterways 3) Vol. 87; August 1961; pp. 1-28.

15. Shu-t'len Li, F.; "Operative Energy Concept in Marine Fendering"; ASCE Proc. (Waterways 1) Vol. 89; February 1963; pp. 37-50.

316

16. Lewis, Edward V. and Borg, Sidney; "Energy Absorption by the Ship";
    Princeton Conference on Berthing in Exposed Locations; pp 69-86.

17. Baker, A.L.L.; "Gravity Fenders"; Princeton Conference on Berthing in
    Exposed Locations; pp. 97-106.

18. Broersma, G. and Middelbiik, C.G.; "Middlebeek Fender, Naval Engineers
    Journal"; April 1971; pp. 83-91.

19. Dent, G.E. and Saurin, B.F.; "Tanker Terminals-Berthing Structures";
    Institute of Civil Engineers: Tankers and Bulk Carrier Terminals.

20. Hopkins, David A.; "The Design of Piers, Jetties and Dolphins", ASCE
    Proc. (Maritime St. #2846) June 1955; pp. 22-38.

21. Little, Donald Hamish; "Some Designf for Flexible Fenders"; ASCE Proc.
    (Maritime St. #21) October 21, 1952; pp. 42-105.

22. Levinton, Zusse; "Elastic Fender Systems for Wharves"; Princeton Con-
    ference on Berthing in Exposed Locations; pp. 87-95.

23. Palmer, Robert W. Blancato, Virgil; "New Retractable Marine Fender
    System"; ASCE Proc. (Waterways #1513) January 1958; p. 1513-1 to
    1513-8.

24. Picco, John and Blancato, Virgil; "Resilient Ferry Slip with Retractable
    Fender System"; August 1968, pp. 297-303.

25. Quinn, A.; Design and Construction of Ports and Marine Structures; McGraw-
    Hill, 3rd Edition; 1976; New York, New York.

26. Stracke, F.H.; "Offshore Mooring Facilities for Tankers up to 100,000
    Dead Weight Ton Capacity"; Princeton Conference on Berthing in
    Exposed Locations, pp1 157-172.

27. "Heavy Duty Fendering"; Tanker and Bulk Carrier, February 1973; pp. 20-22.

28. Leimdorfer, P; "On the Selection of a Pile Type"; Acier/Stahl/Steel;
    September 1971; pp.369-377.

29. "The Fenders Retract"; Wood Preserving News; June 1971; pp. 4-7.

30. Ford, D.B., Young, B.O. and Waler, G.W.; "Hi-Dro Cushion Camel-A New
    Floating Fender Concept"; Conference on Coastal Engineering, London;
    September 1968; pp. 1185-1199.

31. Tam, W.A.; "Dynamic Mooring and Fendering System"; Offshore Technology
    Conference 1971; April 1971; pp. 53-62.

32. Toppler, J.F., Harris, H.R., and Weiersma, J.; "Planning and Design of
    Fixed Berth Stru-tures for 300,000 to 500,000 DWT Tankers"; Offshore
    Technology Conference, May 1972; pp. 279-294.

33. Powell, R.G. and Carle, R.B.; "The Use of Hydraulic Cushioning in the Docking of Super Tankers"; Offshore Technology Conference; May 1972; pp. 769-776.

34. Svendsen, I.A. and Jensen, J.V.; "The Form and Dimension of Fender Front Structures"; Dock and Harbour Authority; June 1970; pp. 65-69.

35. Terrell, M.; "A New Look at Fendering Systems"; Dock and Harbour Authority; May 1972; pp. 9-11.

36. Bijlsma, T.J.; "Super Fenders for Super Tankers"; Dock and Harbour Authority; April 1970; pp. 509-510.

37. Atack, D.C. and Kohring, W.; "Berthing and Mooring Systems for Mammoth Ships"; Dock and Harbour Authority; October 1970, pp. 241-242.

38. Svendsen, I.A.; "measurement of Impact Energies on Fenders"; Dock and Harbour Authority; September 1970; pp. 180-8.

39. Institute of Marine Engineers; "Offshore Mooring Fenders for ULCC's"; Marine Engineers Review; October 1973; P. 97.

40. Analytical Treatment of Problems of Berthing and Mooring Ships, NATO Advanced Study Institute.

41. American Association of Port Authorities, Inc.)AAPA), 1973, Port Planning, Desing and Construction, prepared by Committee IV, Construction & Maintenance, Washington, D.C.

42. Baker, A.L.L., 1953, Paper to the XVIII International Navigation Congress, Rome.

43. Brolsma, J.H., Hirs, J.A., and Langeveld, J.M., 1977, "On Fender Design and Berthing Velocities," 24th International Navigation Congress, Section II, Subject 4, Leningrad, pp. 87-99.

44. Bruun, P., 1976, Port Engineering, Gulf Publishing Co., Houston, Texas.

45. Costa, F.V.m 1964, "The Berthing Ship," The Dock and Harbour Authority, Vol. XLV, May, pp. 22-26, June, pp/ 49-53, July, pp. 90-94.

46. Dent, G.E., and Saurin, B.F., 1969, "Tanker terminals- berthing structures," Conference on Tanker and Bulk Carrier Terminals, The Institution of Civil Engineers, November 13, pp. 49-59.

47. Girgrah, M., 1977, "Practical Aspects of Dock Fender Design," 24th International Navigation Congress, Section II, Subject 4, Leningrad, pp. 5-13.

48. Goldman, J. 1977. Presidents of Friede and Goldman, Naval Architects and Marine Engineers, interviewed December 9.

49. Han, E.H. and Padron, D.V. 1978. Factors Affecting the Design and Construction of Offshore Terminals, Paper presented at meeting of the New York Metropolitan Section, Society of Naval Architects and Marine Engineers, February 15, 45 pages.

50. Horeczko, G. 1978. "Plastic protection for wood piles, "American Seaport, February, p. 24.

51. Lackner, E., and Wirsbitzki, D., 1974, "Fendering for Bulk Carrier Ports and Container Terminals, "Sixth International Harbor Congress, Section 2.19, May 12-18, 9 pp.

52. Lee, T.T., 1965, A Study of Effective Fender Systems for Navy Piers and Wharves, Technical Report R 312 prepared for U.S. Naval Civil Engineering Laboratory, Port Hueneme, Calif., March, 114 pages.

53. Lisnyk, J.A. 1975. "Projected Ship Types to the Year 2000", Ship Structure Symposium Proceedings, October 6-8, Washington D.C., pp. Q1-Q13.

54. Lord Corporation, 1977, Lord Marine Fender Manual, September 15.

55. Marine Engineering/Log. 1976. "LASH takes on lighters and heavyweights in moving NGL plant to Singapore", Vol. 81, No. 6, June, pp. 44 & 45.

56. Marine Engineering/Log. 1976. "U.S. Merchant Shipbuilding, 1607-1976". Vol. 81, No. 9, August, pp. 65-77, 172.

57. Maritime Administration (MarAd). 1965-1977. Merchant Fleets of the World - Oceangoing Steam and Motor Ships of 1,000 Gross Tons and over as of December 31, 19--, U.S. Department of Commerce, publish annually, Washington, D.C.

58. Meyers, J.J., Holm, C.H., and McAllister, R.F., 1969, "Wind and Wave Loads," Handbook of Ocean and Underwater Engineering, McGraw-Hill.

59. Minikin, R.R., 1950 & 1963. Winds, Waves and Maritime Structures, Charles Griffin & Co., London, 1950 (1st Edition) pp. 162-212 and 1963 (2nd Revised Edition) pp. 183-233.

60. Nelson, W.L., 1977. "What are tanker sizes?" The Oil and Gas Journal, Vol. 75, No. 38, September 12, pp. 128 & 129.

61. Nickels, F.J. 1975. "Marine - Port Technology Forecasts and Demand Analyses", Port System Study for the Public Ports of Washington State and Portland Oregon, Volume II, Technical Supplement, Part 5, prepared for Maritime Administration, U.S. Department of Commerce, March 31, p. 117.

62. Oil Companies International Marine Forum (OCIMF), 1977, Prediction of Wind and Current Loads on VLCC's, London

63. Piaseckyj, P.J., 1977, "State-of-the-Art of Fender Design," 24th International Navigation Congress, Section II, Subject 4, Leningrad, pp. 133-143.

64. Quinn, A. DeF., 1961, Design and Construction of Ports and Marine Structures, McGraw-Hill, New York, 531 pages.

65. Rosenblatt, M. & Son, Inc. 1978. Letter dated January 20, in response to future ship size inquiry.

66. Ross, D., 1975. Trends in Merchant Shipping (1969-1980), performed under contract No. N00123-75-C-0403 for the Naval Undersea Center, April 15, 89.

67. Saurin, B.F., 1963, Paper to the Sixth World Petroleum Congress, Frankfurt.

68. Seaward International, Inc., 1977, Sea Cushion Marine Fenders, product catalog SCTM-4(5-77), 51 pages.

69. Seibu Polumer Chemical Co., Ltd., 1977, Seibu Rubber Dock Fenders, manufacturer's catalogue.

70. Stieff, Jr., J.L. and Bayle, J.J., Year Unknown. "Effect of Fungicides on Natural and Synthetic Rubber," Industrial and Engineering Chemistry, Vol. 39, No. 9, pp. 1136-1138.

71. U.S. Senate, Committee on Interior and Insular Affairs, 1974. Deepwater Port Policy Issues, 93rd Congress, 2nd Session, Washington, D.C., 102 pages

72. Vredestein, B.V., 1977. Rubber Dock Fenders, manufacturer's catalog.

INDEX

322

MARINE BORE - 4, 41, 47, 50, 54, 55, 220

MATTE SURFACE - 56

METAL ARMORING - 54

MILL SCALE - 56

MOISTURE ABSORPTION - 60, 63, 70

MOLDS - 40, 41

MOLLUSKS - 47

MOMENT OF INERTIA - 87, 102, 115

MOORING DOLPHIN - 159, 160

MUD LINE - 47, 50

OVERLOAD FAILURES - 40

OXIDATION - 54

PARAMETRIC STUDY - 117

PHOLADIDAE - 47, 50

PIER DOLPHIN - 162

PILOT - 93

PLASTIC - 60, 220

PLATE PANELS - 86

PNEUMATIC FENDER - 3, 31, 37, 100, 107

POPOUT - 70

POROSITY - 66

POWDER-POST BEETLE - 41

POZZOLANS - 82

PRESERVATIVE - 44, 54

PRESTRESSING - 63, 80, 82

RADIUS OF GYRATION - 85, 89

REACTIVE AGGREGATES - 70

VELOCITY - 84, 85, 86, 87, 89, 94, 96, 101, 103, 107, 110, 111, 112, 142

VISCOUS RESISTANCE - 37

VOLTAIC POTENTIAL - 56

WALE - 5, 8, 101, 102, 142, 143, 145, 149, 151

WEEP HOLES - 63

WHITE OAK - 55

WOOD LOUSE - 50

WRAPPINGS - 59, 60

YAW - 85